Journey in Aeronautical Research
A Career at NASA Langley Research Center

by
W. Hewitt Phillips

NASA History Office
Office of Policy and Plans
NASA Headquarters
Washington DC 20546

Monographs in Aerospace History
Number 12
November 1998

William Hewitt Phillips, January 1945, age 26, then head of the
Stability and Control Branch, Flight Research Division.

Table of Contents

List of Figures

CHAPTER 1

Introduction

My career to the present has covered 58 years, all at Langley Research Center in Hampton, Virginia. At the start of my work, the center was called the Langley Memorial Aeronautical Laboratory of the NACA (National Advisory Committee for Aeronautics). With the advent of the space program, it became the Langley Research Center of NASA (National Aeronautics and Space Administration). In subsequent discussions, the name of the center will be abbreviated to simply Langley. During this entire time, my primary interest has been in research in aeronautics and in the related problems of space flight. Actually, I have had two parallel careers because my interest in aeronautics started at a much earlier age, about 9, with the building and flying of model airplanes. This interest has continued through my life. The model airplane hobby has had a powerful influence in contributing to my interest in aeronautical research. These hobby activities will not be mentioned in this account except when they contributed directly to research on full-scale airplanes.

During this period, many advances have occurred in the art and science of aeronautics. My own contributions, like those of other research workers, have largely taken the form of published reports or journal articles. Because of the progress in this field, interest in research done in earlier years can be expected to have declined to the point that these papers are rarely read or even referred to by persons involved in current projects.

The cumulative output of the center and its contributions to the development of the aircraft industry, however, are recognized as being substantial, and an overview of this history has been published (ref. 1.1). The present volume gives a different perspective on the history of Langley by documenting in some detail the experience of my work as an engineer involved largely with the flight testing of full-scale airplanes in the Flight Research Division. This volume may also serve to convey some knowledge of my work to the general public and to later generations of engineers. Though the emphasis in these discussions is on the technical aspects of the work, some autobiographical notes on my own background and education may be of interest.

The contents of this volume have been selected with the following considerations in mind. First, because most of my more important research has been published in the NACA and NASA reports or in technical journals, all of which are readily available in the NASA libraries, no attempt is made to present the technical contents of these reports. Some of the projects reported will be mentioned to indicate their background or importance. Some other technical work, however, was never published, either because of a lack of general interest or because the results were not conclusive. In a few cases, a report that I considered interesting was turned down by the editorial committee. I will include these items in the discussion. In addition, I will discuss some things that I have learned from my research projects that did not appear in published reports. In appendix I, a brief autobiography of my early years, written at the age of 14, is presented. Appendix II contains a complete listing of my professional reports and publications.

My experience at Langley has placed me in a position to discuss not only my personal activities, but to comment on some aspects of the research environment and how it changed through the years. Someone reading the early chapters of this account might wonder how a young engineer could become engaged in so many projects of great interest for both military and civil aviation with much research

equipment and many facilities immediately available and with apparently little supervision of the technical aspects of the work. There is no doubt that the research environment at Langley was favorable to the development of new ideas while keeping a strong focus on the primary objective of aeronautical research. I will insert chapters in the account at appropriate places to comment on this aspect of the research work.

The subjects discussed in this autobiography, though not highly technical, require use of some concepts and nomenclature familiar to aeronautical engineers, but possibly unfamiliar to persons working in other fields. To make the material more readily understood, I will preface some chapters with brief discussions of the background and the terms used in the topic under consideration.

CHAPTER 2

Early Life Through the College Years

Parents and Early Years

I was born on May 31, 1918, in Port Sunlight, Cheshire (now Merseyside), England. This city, a so-called model village across the River Mersey from Liverpool, was built by the Lever Brothers Company as a place for their workers to live. My father, William Phillips, was employed by this company. His background was in chemistry and prior to working at Lever Brothers, he had been employed in several chemists shops (the British name for drug stores) in which he had obtained a good knowledge of pharmaceuticals, cosmetics, and perfumes.

My mother, Bertha Pugh Phillips, prior to her marriage had been headmistress of the Evelyn Street School, a large infants school in Warrington where the students were children in what would, in America, be called grades one through three. Her employment in such a position at the early age of about 28 was most unusual in England at that time and indicates her leadership and educational ability. She had already written articles and a book and gained some note in England in the field of infant education. My parents, I am sure, gave considerable thought as to whether such a promising career should be interrupted by marriage. In those days, the wife was expected to become a homemaker after marriage.

My birth was shortly before the end of World War I, and my mother told how she pushed me in my carriage waving a flag on Armistice Day. Soon after, in 1920, my father was sent to the United States by Lord Leverhulme, founder of Lever Brothers Company, to introduce a new line of cosmetics. Unfortunately, in 1920 came the postwar depression, and the business climate was not right for introducing a new product. My father was given a job at the Lever Brothers Plant in Cambridge, Massachusetts and was given the choice, after three months, of returning to England or staying in America. He elected to stay in America. He returned to England to arrange for the movement of the family and belongings. We all sailed to the United States and resided first in an apartment on Commonwealth Avenue in Boston, Massachusetts. Later we moved to the first floor of a two-family house near the Charles River in Watertown, Massachusetts (figure 2.1).

My father became the chief perfumer at Lever Brothers Company and was responsible for the perfume in the Lever Brothers products such as Lux Toilet Soap, Lifebuoy, and Rinso. Although his work did not involve mechanical knowledge, his hobby was in the field of electrical and mechanical devices. He had acquired a metal-turning lathe (a Drummond lathe, especially intended for hobbyists) while still in England

FIGURE 2.1. Scenes from childhood (clockwise starting in upper left).

(a) With my mother in England, age 9 months.

(b) Riding my car in Watertown, MA. Our house in background.

(c) With my parents and baby sister, Hilda Evelyn Phillips, in Watertown, MA.

(d) On the Charles River, where the steamboat was tested. Our house in background.

and had constructed a steam engine machined from commercially available castings. He acquired this interest from his father, Thomas Phillips, who also did lathe work. Thomas Phillips must have had a remarkable interest and ability in technical matters. He started life as an uneducated coal miner in Yorkshire, but was taught to read by his wife and continued to improve his knowledge through reading and hobbies. My father, before he died, wrote a brief biography of Thomas Phillips which also gives an insight into my father's own interests and early life. After coming to this country, my father made an electric generator to be driven by the steam engine. The combined model is still on display in our home. My father was also a pianist of moderate ability.

Another model made by my father was a steam-powered torpedo boat destroyer that was about four-feet long. One of my earliest memories is the test of this model, when I was five years old, in the Charles River near our house. The model, pulling a piece of string for its retrieval sailed off with unexpected speed and soon became a speck in the distance. The model was pulled back against the full thrust of the engine and shipped water over its low sides. It eventually swamped and sank with a spectacular cloud of steam about 20 feet from shore. The model was brought up from the bottom of the river about a week later, but after that, my father lost interest in the steamboat and it was kept in our basement for many years. I restored the model and equipped it with radio control a number of years ago, before my father's death and it is now on display in my living room (figure 2.2).

Immediately after his test of the steamboat, my father became interested in crystal radio sets, which were just becoming popular about 1923. The construction of successively more complex radios occupied his interest for a number of years.

My recollections of life in Watertown are rather sketchy, but I do recall playing with wind-up trains and learning to balance on a scooter. My sister, Hilda, was born there. The family made a visit to England during the summer of 1923. In 1924, we moved to a single-family house in Belmont, Massachusetts. I started in the public schools there in the second grade at the age of six, having been taught to read by my mother. I was therefore a year younger than my classmates throughout my schooling.

This review of my childhood is not by any means intended as a complete account of my activities, but I will mention a number of things I did that illustrate my interest in aviation and some of the factors that influenced the growth of this interest. The decade of the 1920's was a period of unusual development in aviation with new designs and records for endurance and speed being reported frequently in the newspapers. In Watertown, I had seen the dirigible *Shenandoah* fly over and in Belmont, airplanes like Curtiss Jennies were observed, sometimes performing stunts such as loops and spins. Lindbergh's flight in May 1927 made a deep impression on me, as it did on young people throughout the country. I was particularly influenced by the suspense caused by the length of time Lindbergh was in the air with news of the takeoff coming one morning; sightings of the plane over Newfoundland that evening; the plane appearing over Ireland the following day; and finally, the

FIGURE 2.2. My father's steam-powered torpedo boat destroyer after restoration.

FIGURE 2.3. Our summer cottage at Long Beach, Gloucester, Massachusetts.

landing at Le Bourguet reported in papers the day after that. I had never had any idea that an airplane could stay in the air that long. Somewhat later, when I was in junior high school, the Army Air Corps put on a very large demonstration in which practically its entire fleet of aircraft was in the air at the same time. The sky was filled with formations of 50 to 100 airplanes such as Keystone Bombers and DH-4 biplanes. My diary entry for that day says "782 planes flew over." The sound of scores of Liberty motors droning away simultaneously will never be forgotten.

My first model airplanes were paper gliders with about a 5-inch wing span that were patterned after Lindbergh's airplane. Our family rented a small cottage at Long Beach, near Gloucester, Massachusetts, for about a month in the summer, with my father coming up on weekends (figure 2.3). On rainy and foggy days, I would fly the paper models in the cottage where the high, unfinished room with a peaked roof gave plenty of space for the models to perform stunts. I learned a lot from these models, as I have described in an article (ref. 2.1).

Back in Belmont, we were fortunate to live in a house across the street from the Underwood playground that had a large grassy area with a good slope down from the top, a level area, and then another slope down to the bottom. The playground contained swings, rings for gymnastics, improvised ball fields, and at the bottom, a swimming pool, which was

open all summer. This land had been given to the town by Mr. Underwood, a descendant of one of the earliest families in Belmont and head of the Underwood Deviled Ham company. The pool, established in 1906, was the first outdoor public swimming pool in the United States. I used all these facilities, but the aspects that really helped my hobbies were the ability to glice models down the hill and to sail model boats in the pool (figure 2.4).

At the top of the street lived a boy that I played with whose father, a research doctor at Harvard, also was interested in aeronautical experiments and had a supply of balsa wood. He built solid balsa gliders of two- to three-foot span with long, slender wings and short fuselages. When these models were launched from the top of the playground, they would glide quite a distance. When they were thrown harder, they would do a loop and continue the flight with a series of oscillations. I built a number of these gliders.

My grandmother Phillips in England sent me a pocket line-a-day diary for 1929. As might be expected at the age of 11, my entries in this diary were rather brief, but among other things I did make a note of what models I was building and of the number of airplanes that flew over each day. This diary also started me in the habit, beginning in 1930, of getting 5-year line-a-day diaries that I have kept, with some breaks, throughout my life. These diaries, which recounted mainly my

FIGURE 2.4. Scenes in Belmont, Massachusetts.

(a) Our house (top).

(b) Swimming pool across the street about 1940 (bottom).

FIGURE 2.5. My nonflying scale model airplanes.

(a) Bellanca Skyrocket, made at age 14 (top).

(b) Pitcairn Autogiro, made at age 15 (bottom).

hobby and social activities rather than my professional work, are useful in establishing the dates and sequence of various activities that I would otherwise have forgotten.

I had little guidance in my modeling activities, but I did find in the children's section of the Belmont Public Library a book entitled *Model Airplanes of 1911*. I read this book numerous times, not realizing that the technical material in the book was largely incorrect or that the models described were obsolete. Model airplanes in 1911 were mostly rubber-powered twin pushers that were constructed of spruce and pine and braced with music wire or bamboo. I made a model like those pictured in the book. The propellers were carved from blocks of soft pine obtained by my father from the carpenter shop at Lever Brothers. The wing had a frame of music wire with ribs soldered in and the canard tail was a piece of cardboard curved to a cambered airfoil shape. The model made a very successful flight in the playground and covered about 100 feet in stable flight in a straight line. Naturally, this flight gave me great encouragement and the incentive to build more models.

My interests were not solely confined to models. I participated in all the sports offered by the playground, including sandlot baseball and swimming in the summer and sledding, skiing, and ice hockey in the winter, though I was at this age rather small and never much of an athlete. By the age of 13, I had learned to use my father's lathe. I made a solenoid-operated electric motor (still in existence), a cannon that shot a cork when loaded with a firecracker, and a crude but workable compressed-air motor. Later, as I devoted more effort to model airplanes, not much use was made of the lathe because model airplanes do not require precision machined parts.

In 1929, a large model airplane club, the Jordan—Traveler Junior Aviation League, was started in Boston. The club was sponsored by the Jordan Marsh Company, a department store, and the *Boston Traveler*, a newspaper (ref. 2.2). My mother enrolled me in the club, but I did not immediately take part in the activities because at the age of 11 I was too young to travel into Boston by myself on the street cars and subways. I did, however, get some information on more current model designs and materials, such as balsa wood strips and sheets. I subscribed to the magazine *Model Airplane News*, a McFadden publication, starting with the first issue in June 1929.

On my eleventh birthday I received as a present an Ideal Every Boys Airplane, an ingenious but heavy rubber-powered flying model kit first marketed in 1922. Later, I received a Silver Ace, a potentially better flyer. The Every Boys Airplane flew about 100 feet, like my twin pusher. I got very sick for a few days after building the Silver Ace, probably from inhaling the banana oil fumes in a closed room and I never got it to fly. Soon, however, I was building balsa models of my own design that flew much better. I also built small rubber-powered models with about a seven-inch span that I would fly in the living room. I sold some of these models, now called parlor planes, to my classmates in junior high school for 25 cents a piece. This price was a real bargain since each model had a hand-carved propeller.

By 1932 at the age of 14 I was able to travel into Boston by myself and started to attend regularly the Junior Aviation League meetings and activities. The club held weekly meetings during the winter and monthly contests indoors in winter and outdoors in summer. A building contest was also held each year for a specified nonflying scale model. I built a Bellanca Skyrocket in 1933 and a Pitcairn PA-18 Autogiro in 1934 (figure 2.5).

The autogiro, in particular, was a very complicated scale model subject. These projects were very time-consuming, but I learned a lot about full-scale aircraft construction and about how an autogiro flies. I also built indoor and outdoor rubber-powered models and by 1934, I was competing on even terms with the best flyers in the League. In 1935, I was selected as one of a team of four flyers from the League who were given all-expense-paid trips to the National Model Airplane Contest in Saint Louis, Missouri. This contest is an annual event called by model enthusiasts the Nationals, or Nats for short. The winners, after returning from this trip, were each given one of the newly developed small gasoline model engines for powering model airplanes. I also was on the team to attend the Nats in Detroit in 1936 and 1938. My main accomplishments in these contests were winning second place in the gasoline-powered Texaco event in 1936 with a flight of 30 minutes 12 seconds and first place in the Stout event for indoor stick models in 1938 with a flight of 21 minutes 56 seconds. I was flying in the Senior category for flyers under age 21. The Open Class for flyers over 21 did not exist in the early days of the League. Nowadays in the large contests, almost all the flyers are grown-ups or senior citizens. Young people are now typically more interested in computers and other hobbies than in model airplanes. This lack of interest in model airplanes is due partly to the advanced state of development of models produced by senior citizens through a lifetime of experience. The lack of interest could also be due to urban growth in cities, which eliminated suitable flying sites within a reasonable distance from club activities.

Building and flying model airplanes, particularly indoor models, does involve many technical considerations, almost to the same extent as full-scale airplanes. One of the great incentives for the young people engaging in this hobby was that the design of models was in a stage of rapid development and the young people in their teens could contribute to this development with their own efforts. The record flights of indoor models increased from about 7 minutes around 1928 to 21 minutes in 1938 and has since increased to 30 minutes in 1945 and to 52 minutes in 1979. The latter record stood for 15 years, but was increased to 55 minutes in 1994. The long-awaited goal of a 1-hour

flight was exceeded in 1995 with a flight of 63 minutes 54 seconds. A unique situation existed in the early stages of model development in that the teenagers building and flying the models knew much more about model design and construction than their adult advisors who organized the sport.

The Director of the Junior Aviation League at the time of my participation was Willis C. Brown, a manual arts instructor in the Arlington, Massachusetts schools and an amateur radio hobbyist. He was always interested in the technical aspects of model airplanes and in 1936, he organized a project for the League members to build a wind tunnel for testing indoor model airplanes. At that time, I was attending MIT and became the chief participant in the project (ref. 2.3). The wind tunnel was unique in design and had a diameter of 5 feet and a length of 16 feet with a airspeed ranging from 2.5 to 4 feet per second. This wind tunnel required two years to construct followed by a year devoted to testing and research. I learned much about aerodynamics and instrumentation from this work, particularly since almost all the problems were encountered a year or more before my MIT courses gave information about the same problems.

In my schooling, a few recollections may be mentioned that have a bearing on my subsequent career. In grade school, I was very shy and teachers would have had a hard time detecting any unusual ability. I do have a recollection that by the end of the third grade, I could remember just about everything that had happened in school up to that time, something that I think the average student could not do. This ability started to disappear after that time, however. I was very shy and studious as a child. My marks in grade school were just average, but in the first year of junior high I started to get all A's and this performance continued with a few exceptions through high school and college. Though neither of my parents knew anything about model airplanes, they were always very supportive of my interest in this hobby.

I had a room to use as a workshop and all the supplies required to fulfill my relatively small demands.

The Belmont schools had some excellent teachers. In the class in ancient history in junior high school, I had a project to build a Ballista, an ancient war machine used for bombarding walls. My model was beautifully built and used some parts manufactured on the lathe. It could be cranked up and would fire small stones or blocks of wood.

In high school, the first scientific course was in physics, which I followed easily and once got in trouble with the teacher for trying to correct a mistake that he made. I really appreciated Euclidean geometry in the senior year in high school, which opened up for the first time the methods of scientific logic. One particularly difficult problem was given out by the teacher with no expectation that anyone could solve it. I managed to give a proof of the proposition involving 45 steps. I have always kept this proof and it is reproduced in appendix III. This problem was given in the Mathematical Puzzles section of the MIT magazine *Technology Review* many years later. It can be solved by a much simpler and more elegant proof, but my brute force approach is equally correct.

I graduated from the Belmont High School in 1935. I was salutatorian and prepared an address that the teacher thought lacked interest or inspiration. She encouraged me to write about the subject closest to my heart, aviation. I wrote an essay on this subject in the style of the Sir Roger de Coverley Papers, writings that I had admired in English class, and managed to overcome my stage fright enough to present my carefully rehearsed and memorized talk. The main speaker at the graduation, a Belmont lawyer, gave a talk very similar to the one that I had originally proposed.

College Years

I started at MIT in 1935 while still living at home to save money, but as a result, remained shielded from the social life of the college. In general, the MIT courses were excellent with the exception of the mathematics courses. Perhaps this opinion resulted from my lack of natural ability in abstract reasoning. I was always able to visualize solutions, a useful ability for engineering problems, but generally contrary to the requirements of rigorous mathematics. I had had very little calculus in high school and the problems in the physics course at MIT always required the use of calculus techniques two weeks before they were taught in the mathematics courses. I hope that this scheduling problem has since been corrected at MIT. Later in advanced calculus, the need to define a small quantity ε "no matter how small" was not clear to me, nor was it ever explained by the professors. In the problems encountered, I could visualize what happened as a quantity approached zero.

I was able to do the mathematics problems, but the lack of basic understanding has always prevented me from making much use of the methods of higher mathematics, for example, vector analysis and linear systems theory, that find many applications in aeronautical work. I can recall one problem presented in a physics class that I solved using a method of solution that I had not been taught previously. The problem had to do with the distribution of velocity of a fluid between two parallel plates. The method I used was later taught in the mathematics course in graduate school as the method of undetermined coefficients. The paper was corrected by a graduate student who expected the solution to be obtained by a different method and who did not recognize that I had "invented" a known mathematical technique.

At MIT, most students took the same courses the first two years and did not start to work on their specialty until the junior year. In this year (1937–1938), I took a general theoretical course called Aeronautical Dynamics under Professor Manfred Rauscher. Unlike the later courses which were mostly of a practical nature, Professor Rauscher's course was devoted entirely to the theories of dynamics of rigid bodies and of hydrodynamics. Professor Rauscher was a natural born teacher. He taught the course in a very thorough manner so that no steps in derivations were omitted. As a result, the material and the reasoning behind it were well established in the students' memories. To accommodate the state of the mathematical knowledge of the undergraduates, however, the entire course was taught without the use of vector analysis. A follow-up course to show how the same results could have been obtained with vector analysis techniques would have assisted the students in understanding material encountered later in more advanced textbooks, but I never took such a course or at least not one so clearly presented. In the senior year, courses I took included aircraft structures under Professor Joseph Newell, aerodynamics under Professor Shatswell Ober, stability and control under Professor Otto Koppen, and automotive engineering under Professors E. S. and C. W. Taylor. These professors are mentioned because they taught a whole generation of aeronautical engineers who graduated and entered the aeronautical industry at the time of the tremendous development of aviation that occurred during and after World War II (WW II). These students were very influential in the development of American aviation during this period. I also took a graduate course, Introduction to Theoretical Physics, which gave me a background in dynamics problems that required more advanced techniques than those taught in the undergraduate years. The objective of the Aeronautical course at MIT was to give students a sufficiently broad practical background so that they could design a complete airplane or any subassembly thereof. In addition, a bachelor's thesis was required.

My thesis was on the subject of boundary layers, under Professor Heinreich Peters. The objective was to make measurements of the development of the boundary layer in a spe-

cial boundary-layer tunnel in the basement of the aeronautics building. The pressure gradient down the tunnel could be varied as desired. The data were analyzed to determine the variation of friction drag coefficient in the tunnel, especially during the transition from laminar to turbulent flow. This calculation was performed using the Gruschwitz method, which required graphical integration of the boundary-layer profiles. Professor Peters was also busy designing the Wright Brothers Wind Tunnel at MIT and was rarely available for consultation. In addition, he was a native of Germany and when WW II broke out, he left to cast his lot with the Germans. Nevertheless, the thesis was completed and copied on microfiche (ref. 2.4). In Germany, Professor Peters designed the large pelton-wheel-powered wind tunnels in Modane, France that were taken over by the French after the war and are still in use.

In addition to the required work, I devoted considerable effort to running tests in the Junior Aviation League wind tunnel. In 1939, I took summer employment at the Pratt and Whitney Aircraft Company in East Hartford, Connecticut. I was given a job as a draftsman in the Installation Department. The main project in this group was the installation of the new R-2800 engine in a Vultee YA-19 attack bomber for its first flight tests. This work involved some designing as well as drafting and was finished in time for me to get a flight in the airplane. This engine was later used to power many of the military airplanes in WW II, including the Republic P-47 Thunderbolt fighter and the Consolidated B-24 Liberator bomber. A photograph of the Vultee attack bomber, produced by the Vultee Aircraft Division of the Aviation Manufacturing Corporation with the R-2800 engine installed is shown in figure 2.6.

A comparison with a photograph of the original airplane with a Pratt and Whitney Twin Wasp engine developing 900 horsepower would show little apparent difference. It is remarkable that the R-2800 engine, capable of developing 2000 horsepower (later 2800 horsepower with full supercharging),

FIGURE 2.6. Vultee YA-19 attack bomber with Pratt and Whitney R-2800 engine installed.

would fit in the same cowling as the smaller engine. In the modified configuration, the cowling and engine were moved backward about a foot to aid in balancing the airplane with the heavier engine. The crew of mechanics in the Installation Department stand in front of the airplane in figure 2.7 along with me in shirt and necktie in the right rear.

When I first went to work, I had a talk with John M. Tyler, the vibration expert at Pratt and Whitney. He presented me with a problem concerning design of the engine mounting pads to decouple vertical and pitching oscillations of the engine. I worked on this problem in the evenings during the summer, inasmuch as there was not much other activity to occupy me. By the end of the summer, I had performed quite a lot of analysis, but the final answer appeared to be incorrect. After discussing these results with Mr. Tyler, I put this work aside, but I was always worried about getting the incorrect result. Many years later, after I retired, I got out the problem again and this time obtained the correct answer. The results were published in a NASA Technical Memorandum (ref. 2.5).

Unfortunately, by this time, the interest in the results had disappeared because the analysis applied to radial piston engines, which have been replaced by gas turbine engines on most high-speed airplanes.

I was involved in several other projects at Pratt and Whitney, including design of an installation for a proposed in-line vertically opposed engine, design of an installation of an R-2800 in a British Vickers Wellington bomber, and preparation of an exhibit of a futuristic engine installation for the New York 1939 Worlds Fair. Most all of these projects were behind schedule and had to be finished during the summer.

After my work at Pratt and Whitney, I concluded that I was not suited for the work at an industrial concern, with its rushing to meet deadlines and its lack of time to study problems in depth. At MIT, I had become acquainted with the research work at the NACA and decided to try to work there. Before leaving MIT, however, I stayed an additional year to obtain a master's degree.

The graduate year included aeronautical courses of a more theoretical nature. In addi-

FIGURE 2.7. Mechanics in the Installation Department of Pratt and Whitney in front of YA-19 airplane. I am at right rear.

tion, there was a course in instrumentation by Professor Charles Draper. Dr. Draper was already noted for his work on aircraft and engine instrumentation. Later, he became famous for the invention of the inertial navigation system and for his work on gyroscopic gun directors. This course perhaps more than any other helped prepare me for research work because the subjects he taught were those that he was working on as research problems. He also brought in some aspects of electronics and physics as well as aerodynamics.

The master's degree also required a thesis. I wrote a thesis *Exhaust Gas and Radiator Propulsion* under Professor Rauscher, who was also unavailable for much consultation. This thesis could have provided an inspiration for the then unknown jet engine, but I was discouraged from considering such developments by statements made in earlier courses that metals could not withstand the high temperatures required. This thesis was later published, with small changes, in the *Journal of the Aeronautical Sciences* (ref. 2.6).

An option in the graduate year was to perform an independent research project in addition to the thesis. I was still interested in boundary layers and made studies of the effect of air velocity on an electric arc with the object of using these effects to measure air flow in the boundary layer. I was again entirely on my own and obtained a large 30,000 volt transformer from the electrical engineering storeroom to produce the arc. The voltage drop across the AC arc was measured with an electrostatic voltmeter, which inherently rectifies the AC voltage. This was a bulky and delicate instrument, which was also borrowed from the electrical engineering storeroom. I was afraid of damaging the expensive instrument. I therefore built one of my own that worked on the same principle. My electrostatic voltmeter had attracting plates and chambers made from tin cans and a suspension constructed from the tungsten wire used for bracing indoor models in place of the quartz fiber in the professional instrument. This constructed instrument worked as well as the professionally built one. The results were put out in a paper that received a high grade from Professor Draper, but

FIGURE 2.8. Model glider equipped with phugoid damper.

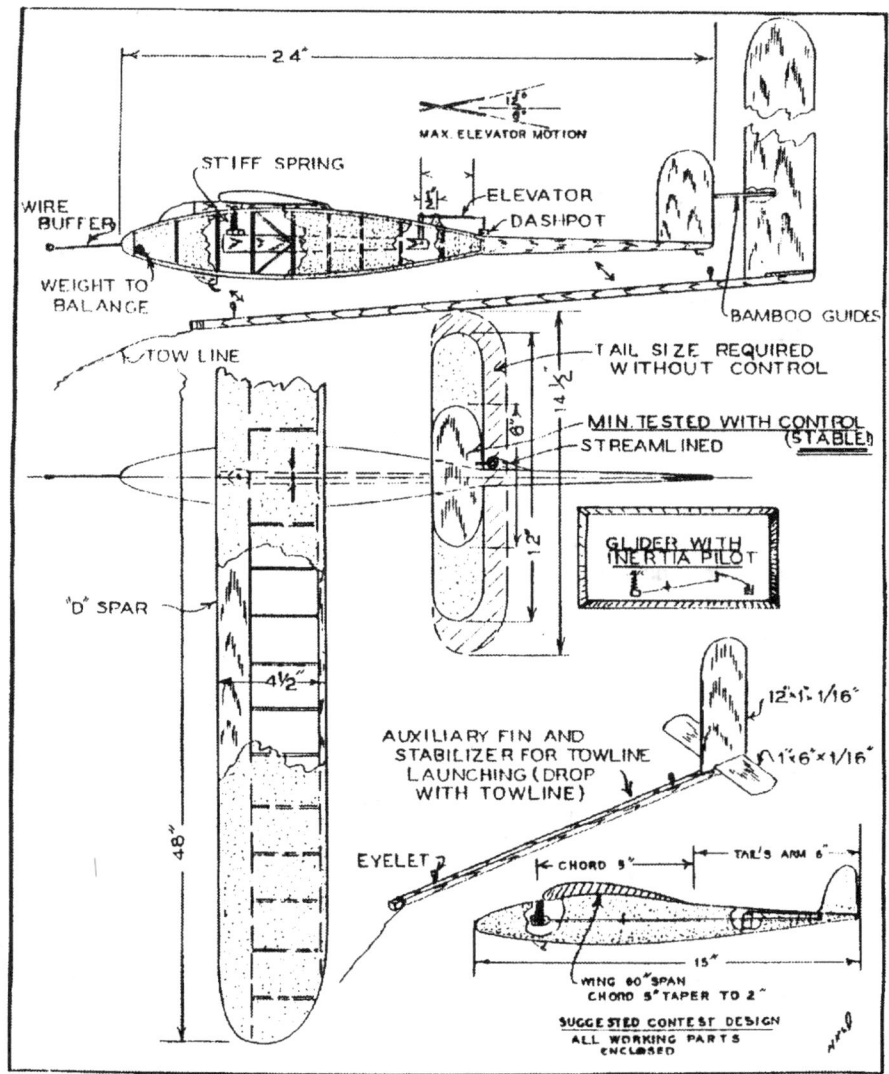

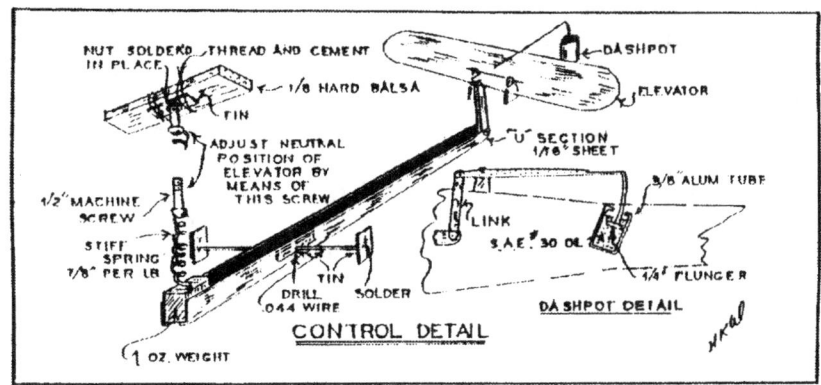

apparently has since been lost. I was fortunate, probably, to escape electrocution in this project because the transformer stores a large amount of energy and could easily kill a person. I did receive one pretty sharp shock from it. I was quite naive in handling high voltage equipment and should not have been allowed to work alone with it.

I also became interested in automatic control of airplanes. The theoretical methods for analyzing such systems were just being developed and were presented in theses by two other MIT graduate students, Herbert K. Weiss and Shih-Nge Lin. I knew from my childhood model glider experience that models with short fuselages and small horizontal tails would have a poorly damped longitudinal oscillation (the so-called phugoid oscillation). I devised a method using a spring-mounted weight and a viscous damper to operate the stabilizer in a way that I thought would damp out this oscillation. I made an analysis of the system that indicated favorable results. The system was then installed in a model towline glider of about a 40-inch wing span with a very tiny stabilizer. When I tried the model, it was difficult to get convincing results because of the difficulty of towing a glider up and launching it in a wind in a consistent attitude.

I did, however, feel confident enough to give a demonstration to Professor Koppen on the MIT athletic field. The model performed perfectly. On the first launch, with the stabilizer locked, the model went into a continuous phugoid oscillation. On the next flight, with the system operating, the oscillation damped out immediately and the model made a smooth glide. These results showed the advantages of using model airplanes to study stability problems. Today, such tests are called dynamic model tests and utilize modern equipment such as radio control and tele-metering to obtain the data. In 1940, however,

such tests were quite rare. Later, a drawing of my phugoid damper and an article describing its operation was prepared by Herbert K. Weiss. The article was submitted to the editor of a modeling publication, but so far as is known, was never published. A copy of Weiss' drawing of the phugoid damper is given in figure 2.8. Incidentally, I have kept in touch with and often obtained advice from Mr. Weiss throughout my career. He is a brilliant engineer who became noted for his work in automatic control, missile guidance, and operations research.

My interest in the scientific aspects of model aviation continued with a series of tests exploring the drag of fuselages, which I made in an old MIT wind tunnel that dated from the early 1920's. The tunnel and its balance were copies of the early wind tunnel in the National Physical Laboratories in Teddington, England. The tunnel was long outdated for research on full-scale airplanes, but it was quite suitable for tests on outdoor, gasoline-powered model airplanes. An article publishing these results was presented in the magazine *Model Airplane News* and recently in the Twenty-Seventh Annual Symposium of the National Free Flight Society (ref. 2.7).

During the graduate year at MIT, I went to Boston to take the Civil Service Exam for Junior Aeronautical Engineer. Though I was anxious to work for the NACA, they were not overly anxious to have me. The exam was quite tough. I later learned that the questions had been supplied by Eastman Jacobs, perhaps the most noted aerodynamicist at Langley. I waited a long time to get the results, but just as I was starting to think about looking into other job opportunities, I received a notice to report for duty. I entered duty at the NACA at the Langley Memorial Laboratory on July 1, 1940 and was assigned to the Flight Research Division.

CHAPTER 3

Setting for the Research Work at Langley

It is well known to those who worked at the NACA in its early years that little attempt was made to publicize the activities of the organization outside the circle of interested parties in the military services and the aeronautical industry. As a result, the general public was largely unaware of the existence of the Langley Memorial Aeronautical Laboratory or of the work conducted there. Inasmuch as all the work described herein was conducted in this location and setting, a brief description of the origin and history of the center may be useful in understanding the development and progress of the organization.

History

The NACA, or National Advisory Committee for Aeronautics, was established by Congress in 1915 to encourage the development of aviation in the United States. From the standpoints of both research and industrial development, progress in aviation in this country was far behind that in the warring nations of Europe. The organization established by Congress to correct this situation was an independent government agency reporting directly to the President. The main governing body of the organization, often referred to as the Main Committee of the NACA, consisted of about 20 men representing the military services, the aircraft indus-

try, and universities and research organizations. These officials served without compensation and met two or three times a year. One of the specified objectives of the committee was to establish a center for aeronautical research.

The Langley Memorial Aeronautical Laboratory was established in 1917 on Langley Field, already a base for the Army Air Corps that had been established in April 1917. Langley Field, now Langley Air Force Base, is located in Hampton, Virginia. Hampton, in turn, is on the Virginia Peninsula, which is a strip of land bounded by the James and York Rivers that extends down to the entrance of the Chesapeake Bay.

By 1920, the first research facilities were in place and aeronautical research was started with an initial complement of 4 professionals and 11 technicians. Before describing the later developments, some information on the features of this location are presented.

Description of Surroundings

Many historical areas are located on the Peninsula, including Jamestown, the first English settlement in the United States, which dates from 1607. Hampton is known as the first permanent English settlement in the United States. Other nearby towns dating from colonial times are Williamsburg and Yorktown.

When I arrived at Hampton in 1940, it was a town of about 7000 people. The main industries of the local population were fishing and crabbing. Many more people were located in the surrounding area of Elizabeth City County. Hampton was later incorporated as a city taking in the surrounding area. The other large city on the Peninsula was Newport News, adjacent to Hampton. It had a much larger population at that time. Its principal industry was the Newport News Shipbuilding and Drydock Company.

Langley Field was selected as the site of the research center for the NACA because of the availability of the airfield; the distance from Washington was large enough to avoid political interference, but close enough to allow convenient travel; and the weather was suitable for flight research. Records had shown that Langley Field had more clear days per year than any other base on the East Coast.

The early 1940's included the occurrence of WW II, a period of tremendous expansion at Langley and of the aircraft industry in general. I have prepared an article, "Recollections of Langley Memorial Aero Lab in the Forties," that describes the activities during this period (ref. 3.1). A summary of some aspects of the social life and the surroundings at Langley and in the town of Hampton, Virginia, where it is located, may be helpful in setting the stage for the type of work done during the subsequent years.

In 1940, the Langley Memorial Aeronautical Laboratory was still quite a small center with fewer than 750 employees. Early employees at the Laboratory have told what an isolated and provincial place Hampton was when the NACA first started operations in the early 1920's. The first few engineers met with considerable animosity from the local population and the lack of any social or cultural activities made life difficult, especially for the wives. By 1940, when I arrived, these conditions had largely disappeared. I found

life extremely happy and exciting, particularly since this was my first experience living away from home and without the rigorous routine of studying at MIT. The small size of the Laboratory meant that most of the employees knew a good number of the staff personally and the small social groups and organizations that had formed made for a pleasant social life as well as providing for a good interchange of research ideas.

To a large extent, the employees made their own social activities. The only professional society, the Engineers Club, was founded in 1940 and put on excellent programs. There was a very active model airplane club, an item of importance to me, since this had been my main hobby before coming to Langley. The NACA hired about 100 expert model builders, starting about 1938. A special Civil Service Exam for which the only requirement was that the applicant should have won a model airplane contest was used to select the model builders. Those selected were employed in model shops or instrument shops, and later many of them rose to high positions at Langley. After arriving here, these model builders kept up their interest in model airplanes by holding regular club meetings and large contests. Langley engineers gave advice to the modelers and officials took an active part in sponsoring and awarding prizes at contests.

The NACA Tennis Club had been started in the 1920's and had six clay courts made by rolling the local soil, which proved very satisfactory for the purpose.

A social organization called the "Noble Order of the Green Cow" put on dances for the employees. These dances were often held at the Hampton Armory and were very well attended.

During the 1930's, quite a few engineers flew their own light planes. This activity, of course, ended with the start of the war except for the Civil Air Patrol. There were at least three private airports in the nearby cities of Hampton and Newport News.

Because of the location on a peninsula, water sports were popular. Many NACA engineers became members of the Hampton Yacht Club and engaged in the club activities.

Golf was also available with clubs at Hampton and Yorktown. The Yorktown golf course was quite interesting because it was located on a Revolutionary War battlefield. Later, this course was closed because the Secretary of the Interior, Harold Ickes, ruled that this was not a suitable activity for a national monument. I felt that I learned more about Revolutionary War history by playing golf among the fortifications and redoubts, however, than by reading books on the subject.

This seemingly ideal situation at Langley was interrupted by WW II. With the entrance of the country into the war following the attack on Pearl Harbor in December 1941, the work at the Laboratory took on a new urgency. The work week was increased from 44 to 48 hours, which required a full day of work on Saturdays. In addition, the imposition of gas rationing prevented much travel. Everyone worked very hard during the war. The increased work load also required a tremendous expansion in facilities and personnel. Before describing this expansion, however, I will review some of my first research assignments at Langley.

First Assignments in Flight Research

In my previous acquaintance with the work at Langley, I had learned that most of the research was performed in wind tunnels. Certainly my previous experience, particularly with the Junior Aviation League wind tunnel and the boundary-layer research of my bachelor's thesis, would have ideally prepared me for wind-tunnel work. Nevertheless, I gladly accepted the assignment to the Flight Research Division. The decision to place me in this division certainly had a profound effect on my subsequent career. Most

of the wind tunnels have a specific purpose and range of airspeed. A person assigned to a wind tunnel would naturally become a specialist in the particular research objectives of that facility and, if so inclined, would contribute to the theory involved in that phase of research. Flight research, however, involves all types of aeronautical problems. The research engineer must have a knowledge of such diverse fields as structures, aerodynamics, loads, performance, and stability and control. As a result, I became a generalist, with less than complete knowledge in most fields, but better acquainted than most young engineers with the research work being performed at all the facilities at the Laboratory.

The working conditions were also very desirable, both in terms of personnel and facilities. The head of the division when I started work was John W. (Gus) Crowley, Jr. who had performed much early flight research and later became Chief of Aerodynamics at NASA Headquarters. Second in command was Floyd L. Thompson, who became Director of Research and later held the positions of Associate Director and Director of Langley. The division was divided into sections entitled Aircraft Loads, Performance, Helicopters, and Flight Research Maneuvers. The latter title was later changed to Stability and Control. I was assigned to the Flight Research Maneuvers Section under Robert R. Gilruth, who had just started work three years earlier, but was already recognized for his outstanding ability to direct research. He later became director of the Johnson Space Center during the Mercury, Gemini, and Apollo programs and the initial stages of the Space Shuttle program.

The facility available for flight research consisted of the Flight Research Hangar, stocked with about 15 to 20 airplanes of all types, such as light airplanes, transports, and the latest military airplanes including both fighters and bombers. Mechanics were available to maintain the airplanes and to install special research equipment. Recording instrumentation had been developed from the first

FIGURE 3.1. Airplanes being tested at the NACA Flight Research Hangar in 1944.

days of the NACA and was operated by technicians skilled in its use and installation.

These airplanes were flown by test pilots. The job of the Flight Research Engineer was to plan the flight tests, analyze the instrument data, and prepare reports on the results. I particularly enjoyed working in the offices that were located on the airfield side of the hangar, which allowed a clear view of the take-offs and landings of all the military and research airplanes. Later, however, an additional hangar was built and the offices were moved to the other side.

A photograph of the airplanes that were tested in the Flight Research Division in 1944 is shown in figure 3.1. This photograph

was taken after the construction of the second hangar.

With this background, I will proceed to describe some of the early work that I did at Langley. I will not attempt to give a chronological account of my activities, which would become confusing because of the overlapping of many jobs. Instead, I will single out some of the more important categories of research and collect the activities that fall under these headings. To make the discussions more understandable to persons not trained in aeronautical engineering, I start with a brief discussion of the history of the discipline and definitions of the technical terms used.

CHAPTER 4

Flying Qualities

By "flying qualities" are meant the stability and control characteristics of an airplane that have an important bearing on the safety of flight and on the ease of controlling an airplane in steady flight and in maneuvers. The term "handling qualities," which is applicable to other types of vehicles is used synonymously with flying qualities when applied to airplanes.

To start the discussion of flying qualities, the concept of stability should be understood. Stability can be defined only when the vehicle is in trim; that is, there are no unbalanced forces or moments acting on the vehicle to cause it to deviate from steady flight. If this condition exists, and if the vehicle is disturbed, stability refers to the tendency of the vehicle to return to the trimmed condition. If the vehicle initially tends to return to a trimmed condition, it is said to be statically stable. If it continues to approach the trimmed condition without overshooting, the motion is called a subsidence. If the motion causes the vehicle to overshoot the trimmed condition, it may oscillate back and forth. If this oscillation damps out, the motion is called a damped oscillation and the vehicle is said to be dynamically stable. On the other hand, if the motion increases in amplitude, the vehicle is said to be dynamically unstable.

The theory of stability of airplanes was worked out by G. H. Bryan in England in 1904 (ref. 4.1). This theory is essentially equivalent to the theory taught to aeronautical students today and was a remarkable intellectual achievement considering that at the time Bryan developed the theory, he had not even heard of the Wright brothers' first flight. Because of the complication of the theory and the tedious computations required in its use, it was rarely applied by airplane designers. Obviously, to fly successfully, pilotless airplanes had to be dynamically stable. The airplane flown by the Wright brothers, and most airplanes flown thereafter, were not stable, but by trial and error, designers developed some airplanes that had satisfactory flying qualities. Many other airplanes, however, had poor flying qualities, which sometimes resulted in crashes.

Bryan showed that the stability characteristics of airplanes could be separated into longitudinal and lateral groups with the corresponding motions called modes of motion. These modes of motion were either aperiodic, which means that the airplane steadily approaches or diverges from a trimmed condition, or oscillatory, which means that the airplane oscillates about the trim condition. The longitudinal modes of a statically stable airplane following a disturbance were shown to consist of a long-period oscillation called the phugoid oscillation, usually with a period in seconds about one-quarter of the airspeed in miles per hour and a short-period oscillation with a period of only a few seconds. The lateral motion had three modes of motion: an aperiodic mode called the spiral mode that could be a divergence or subsidence, a heavily damped aperiodic mode called the roll subsidence, and a short-period oscillation, usually poorly damped, called the Dutch roll mode.

Some early airplane designers attempted to make airplanes that were dynamically stable, but it was found that the requirements for stability conflicted with those for satisfactory flying qualities. Meanwhile, no information was available to guide the designer as to just what characteristics should be incorporated to provide satisfactory flying qualities.

By the 1930's, there was a general feeling that airplanes should be dynamically stable, but some aeronautical engineers were starting to recognize the conflict between the requirements for stability and flying qualities. To resolve this question, Edward P. Warner, who was working as a consultant to the Douglas Aircraft Company on the design of the DC-4, a large four-engine transport airplane, made the first effort in the United States to write a set of requirements for satisfactory flying qualities. Dr. Warner, a member of the main committee of the NACA, also requested that a flight study be made to determine the flying qualities of an airplane along the lines of the suggested requirements. This study was conducted by Hartley A. Soulé of Langley. Entitled Preliminary Investigation of the Flying Qualities of Airplanes, Soulé's report showed several areas in which the suggested requirements needed revision and showed the need for more research on other types of airplanes (ref. 4.2). As a result, a program was started by Robert R. Gilruth with Melvin N. Gough as the chief test pilot. It was during the course of this program that I entered the Flight Research Division and my early work consisted largely of flying qualities investigations.

The technique for the study of flying qualities requirements used by Gilruth was first to install instruments to record relevant quantities such as control positions and forces, airplane angular velocities, linear accelerations, airspeed, and altitude. Then a program of specified flight conditions and maneuvers was flown by an experienced test pilot. After the flight, data were transcribed from the records and the results were correlated with pilot opinion. This approach would be considered routine today, but it was a notable original contribution by Gilruth that took advantage of the flight recording instruments already available at Langley and the variety of airplanes available for tests under comparable conditions.

An important quantity in handling qualities measurements in turns or pull-ups is the variation of control force on the control stick or wheel with the value of acceleration normal to the flight direction expressed in g units. This quantity is usually called the force per g. This notation will be used in this monograph.

Flying Qualities Tests

By the time I started work, Mr. Gilruth had tested about 16 airplanes. These airplanes ranged from light airplanes to the largest airplane then available, the 150-foot span Boeing XB-15 bomber. Based on these tests, Gilruth prepared a report, Requirements for Satisfactory Flying Qualities of Airplanes, that was published as a NACA Technical Report in 1943, but was available in preliminary form when I arrived at Langley (ref. 4.3). This report served as the basis for much of the work of the Stability and Control Section during the ensuing years. After a period of indoctrination, I was the engineer in charge of the flying qualities measurements. Pictures of me with two of the airplanes tested are shown in figure 4.1.

My name first appeared on reports on the flying qualities measurements of the Vought-Sikorsky XF4U-1 Corsair airplane, a Navy Fighter well-known for its service in WW II (figure 4.2).

FIGURE 4.1. Pictures of me with two of the airplanes tested in my first years at the NACA.

(a) Curtiss P-40 fighter (top).

(b) Brewster XSBA-1 scout bomber (bottom).

This study had been started before I arrived and my main contribution was to edit the text. The first published NACA report for which I supervised the tests (May 1941) was on the flying qualities measurements of a Curtiss P-40 airplane (ref. 4.4). This airplane was the primary U.S. fighter at the start of WW II (figure 4.3). Need for some improvement on the handling qualities of this airplane was shown by several ground-looping accidents directly in front of my office windows as the Air Force squadrons came in for landings.

After the start of WW II, the famous British fighter airplanes, the Hawker Hurricane, (figure 4.4) and the Supermarine Spitfire (figure 4.5) were obtained for tests at Langley.

Brief summaries of some of the measured flying qualities of interest are included in the figure captions of these and other airplanes.

FIGURE 4.2. Vought-Sikorsky XF4U-1 airplane. The flying qualities of this airplane are of interest as one that was in service before the publication of Gilruth's flying qualities requirements. The elevator control forces in turns were found to be desirably light, but the aileron forces in high-speed rolls were heavy, resulting in sluggish response. The rudder force variation with speed for trim in high-speed dives and strafing runs was large, resulting in difficulty in holding the sights on an aim point. Control forces in the carrier approach condition had an unstable variation with speed, a common condition that does not have a very adverse effect on the flying qualities.

FIGURE 4.3. Curtiss P-40 airplane. This airplane at the start of the war was lightly armed and underpowered compared to British and German fighters. Its flying qualities were satisfactory except for the usual heavy aileron control forces. Stalling characteristics were poor in some conditions. The airplane had a particularly bad tendency to ground loop, which was found to be caused by asymmetrical stalling of the wing in the three-point attitude. This problem was cured by extending the tail-wheel strut so that the airplane remained unstalled on the ground.

FIGURE 4.4. Hawker Hurricane airplane. A heavily armed fighter airplane noted for its role in the Battle of Britain, the Hurricane's flying qualities were found to be generally satisfactory. The most notable deficiencies were heavy aileron forces at high speeds and large friction in the controls.

FIGURE 4.5. Supermarine Spitfire airplane. A high-performance fighter noted for its role in the Battle of Britain and throughout WW II, the Spitfire had desirably light elevator control forces in maneuvers and near neutral longitudinal stability. Its greatest deficiency from the combat standpoint was heavy aileron forces and sluggish roll response at high speeds.

I published reports on the Hawker Hurricane (April 1942) (ref. 4.5) followed shortly by one on the Spitfire. The data obtained in these tests served to confirm most of the requirements previously proposed by Gilruth. Other reports followed comparing these results with published data on the German fighter Me109 and with U.S. fighter airplanes. During the war, pilots' lives depended on small differences in performance between the first-line fighters, and

continual detailed improvements were made in these fighters. Several research studies were made on improvements, usually on control systems, and close contact was kept with the manufacturers through conferences and preliminary reports.

The tests on the high-speed fighters confirmed the findings of Gilruth that though all the airplanes exhibited instability in the spiral and phugoid modes of motion, these modes did not concern the pilot because his normal control actions prevented the modes from developing to a point that they were noticeable. That is, the airplanes were spirally unstable, but the rate of divergence was small enough that it was not discernible to the pilots. Also, the long-period longitudinal mode might have been a slow divergence or a poorly damped or unstable phugoid oscillation, but the divergence was so slow or the oscillation had such a long period that it was not noticeable in normal flight. The short-period lateral oscillatory mode, the Dutch roll, was noticeable but adequately damped and the short-period longitudinal mode was so well damped that it could not be detected by the pilots. In general, these results applied to most airplanes of this period and explain why successful airplanes could be built without the need to consider theoretical predictions of dynamic stability. On the other hand, Gilruth had found that many of the quantities that could be determined without the need for complex theories, such as control deflections and control forces required in straight flight and maneuvers, trim changes due to power and flap setting, limits of rolling moment due to sideslip, and adequacy of the control effectiveness in maneuvers, were extremely important to the pilot. The tests on the fighter airplanes showed that the longitudinal control force gradient in maneuvers, known as the force per *g*, was a very important quantity, whereas the control force and position variation with speed in straight flight was of less importance and mainly influenced pilot fatigue on long flights. These airplanes were found to be quite satisfactory in most respects, though the aileron effectiveness at high speeds was low because of the large control force required to deflect the ailerons, which was an adverse characteristic in air combat. The detailed improvements mentioned previously were mainly directed at this aileron effectiveness problem.

In addition to analyzing the flight test data, I did theoretical analyses showing the relation between the airplane design characteristics,

such as the center-of-gravity location, tail size, and control surface design, and the resulting flying qualities.

Stall Tests

Stalling characteristics are an important consideration in studying flying qualities because many accidents have resulted from poor stalling characteristics. The design features leading to poor stalling characteristics were at that time, and to some extent still are, poorly understood because stalling is a complex problem involving separated flow. Separated flow occurs when the thin layer of slowly moving air near the surface of the wing, called the boundary layer, thickens and causes large changes in the external flow.

My first assignment at Langley was to analyze data and plot figures from stall tests of the North American BT-9B airplane, a trainer that had displayed dangerous stalling characteristics (figure 4.6).

The flight studies, which had been started before I arrived at Langley, were quite detailed and included tuft studies, flow field surveys in the vicinity of the tail, and tests of slats covering various portions of the span. By way of explanation, tuft studies refer to tests made with numerous short pieces of yarn attached to the wing or tail to show the direction of flow at the surface. If the flow at the surface is reversed, a stalled condition is indicated. Flow field studies required rakes holding a series of flow-direction and velocity sensors to determine the flow characteristics ahead of the tail. Slats are small airfoils mounted ahead of the leading edge of a surface to prevent flow separation at the leading edge and increase the angle of attack that results in a stall. The project engineer on these tests was Maurice D. White, an engineer who later moved to the Ames Research Center and who was responsible for much notable work in the stability and control of airplanes. Perhaps because of the large amount of data, it was impossible to draw any simple conclusions. In most NACA

FIGURE 4.6. North American BT-9B airplane. This airplane had displayed violent roll-off at the stall with no warning, a particularly unsatisfactory condition in a trainer.

reports of that time, the desire was to present a straightforward set of conclusions. As a result, this report was never published, but I have always kept a copy of the rough draft. Through the years, when questions arose concerning the stalling characteristics of airplanes, it was usually possible to find some applicable information in the BT-9 data.

Brewster XSBA-1 Airplane

A project using the Brewster XSBA-1 airplane was started some years before I came to work at Langley. In this designation, X stands for experimental, SB for Scout Bomber, and A for the Brewster Aircraft Company in Long Island, New York, which failed due to labor troubles and was taken over by the Naval Aircraft Factory. This project, a joint effort by the Navy and the NACA, involved testing a series of horizontal and vertical tail surfaces of different sizes and different ratios of control surface chord to main surface chord. Also, shims were provided to attach the wing panels to the fuselage at different values of dihedral angle. Though I am not familiar with the thinking

that led to this project, the objective was probably to supply designers with systematic design data for obtaining desirable flying qualities on new Navy airplanes.

The airplane was quite up-to-date at the time it was designed in the 1930's, but rapidly became obsolete with the advance of military aircraft design. As a result, the studies conducted were of little interest for this particular airplane, but were considered to have general research interest. I was given the job of performing research on the effects on flying qualities of the different configurations (figure 4.1). The project obviously generated much less enthusiasm than tests of the latest fighter airplanes, but I dutifully ran tests on the original configuration and on the first tail modification and wrote reports on the results. Later two other combinations were tested, one with a different vertical tail and one with increased dihedral. To some extent, these projects were used for education of some of the new engineers. The last report was not published until 1946.

The entire project on the XSBA-1 is an excellent example of a study that is unsuited for flight research. The design and construction of flight-worthy components is a lengthy process and the installation of these parts on an airplane in the field is also very time-consuming. The number of possible combinations to be tested is excessive, and if a thorough study of each were made, the project would last a lifetime. The items chosen for test were not the most critical for obtaining good handling qualities and the effect of the configuration changes could have been predicted with sufficient accuracy from available knowledge. Some effects, such as the effect of dihedral, were known even to the Wright brothers. Had complete tests been conducted, the data would not necessarily be applicable to airplanes of different design.

The method chosen by Gilruth of making detailed studies on many available airplanes of different designs proved to be much more effective in yielding new knowledge that is applicable to future designs. This method has been used by all the NACA Research Centers, the military services, and industrial organizations ever since his early work. Flying qualities studies continued at Langley until the start to the space program in 1958. In chapter 16 of this volume, an account is given of the last test made at Langley on a high-speed fighter airplane, the Vought F8U-1. In 1988, I wrote a summary report, *Flying Qualities from Early Airplanes to the Space Shuttle*, that was presented as the Dryden Lecture of the American Institute of Aeronautics and Astronautics (ref. 4.6).

Short-Period Longitudinal Oscillations

The short-period longitudinal oscillations of all airplanes of the period prior to WW II were well damped. The only cases encountered in which this motion was considered unsatisfactory involved coupling between the elevator motion and the airplane motion with the control free. This problem had been observed on the Lockheed 14A transport airplane on which the elevators were not mass balanced (figure 4.7). This type of oscillation could become quite violent. As a result, engineers at Langley were working on methods to predict these problems.

A study in progress when I arrived at Langley was measurement of longitudinal oscillations with free elevator control for a Fairchild XR2K-1 airplane (Navy designation for the Fairchild 22, a parasol-wing light plane). Robert T. Jones and his coworkers in the Stability Research Division had published a series of reports on the effect of free controls on the stability of an airplane, thereby extending the existing theory about the stability with controls fixed. To check this theory, William Gracey in the Flight Research Division had modified the longitudinal control system of the XR2K-1 to include a link with movable weights so that the inertia of the longitudinal control system could be varied. This system was closely related to the device that I had tried on the towline glider at MIT. I discussed the analysis that I had made at MIT. As a result, I became acquainted with Robert T. Jones and was given some responsibility in running the tests. One highlight of the tests was to measure the damping of the elevator oscillations in the Langley Full-Scale Tunnel where the airplane had been mounted for other studies. Motion of the elevator was recorded when the control stick was deflected and released. It was a bitterly cold day when the tests were made with the temperature in the tunnel near freezing. I climbed up a ladder into the open cockpit bundled up in an overcoat, which provided little protection when the airspeed was increased to 100 miles per hour. I served as the "pilot" to deflect the control stick abruptly and record the motions of the elevator. In my entire experience at Langley, I never had the opportunity to run another wind-tunnel test until after I had retired. The Fairchild tests correlated poorly with theory because of excessive friction in the control system, but the results were nevertheless published as an Advance Restricted Report (ref. 4.7).

FIGURE 4.7. Lockheed 14H transport airplane, a later model of one of the original 16 airplanes on which Gilruth's flying qualities requirements were based. This airplane exhibited a coupling between the longitudinal oscillation of the airplane and the motion of the mass unbalanced elevator, which proved very objectionable in rough air. This experience resulted in a requirement to rule out such oscillations, which may result from a variety of causes.

Spiral Instability

Almost all full-scale airplanes exhibit an instability in which the airplane, when the controls are held fixed in level flight, gradually veers off into a diving turn or spiral. This instability, called spiral instability, had been known since the earliest days of aviation, but was generally ignored because instinctive control inputs by the pilot correct the condition so readily that he is rarely aware of its existence. To apply these control inputs, however, the pilot must have a reference with which to judge the attitude, either from the horizon or from cockpit instruments. Without such a reference, the pilot in an airplane is completely unable to determine the attitude, because the familiar vertical reference supplied by gravity is combined with forces in other directions caused by the accelerations of the airplane. Student pilots or other pilots, who are not sufficiently proficient in instrument flying, frequently get in to such a spiral when they fly into a cloud or fog bank. The maneuver has been called the "graveyard spiral" because so many pilots have been killed by it.

Despite many efforts over the years to make this motion of the airplane inherently stable, no solution had been obtained that did not have other serious unacceptable consequences. Gilruth suggested a possible solution consisting of a downward-deflected fixed tab on each aileron. These tabs would tend to make the ailerons float up. Because the ailerons are interconnected through the control system, they would not float up, except possibly a small amount due to flexibility of the control system. In a turn, however, the outboard tip would be traveling faster and the aileron on that tip would have greater tendency to float up than the one on the inboard tip. The result would be to cause the aileron system to deflect in a direction to oppose the turn.

I undertook the job of analyzing this arrangement. At first, the idea seemed promising because for the example chosen, the aileron motion would be sufficient to make the airplane spirally stable up to a lift coefficient value of 1.5, which corresponds to a low-speed flight condition in which the spiral stability is normally most severe. In an actual airplane, however, the aileron movement is opposed by friction in the control system. I then calculated how steeply the airplane would have to turn to cause the ailerons to start to move against the control system friction. Using the measured aileron friction in the Curtiss P-40 fighter airplane, which amounted to 1.5 foot-pounds per aileron at any reasonable value of airspeed, the results showed that the ailerons would never develop enough hinge moment to overcome the static friction. This result is typical for all such schemes that attempt to eliminate spiral stability. The motions are so slow that the aerodynamic hinge moments involved are small, and any friction moments are sufficient to overshadow them. Even on an airplane that is inherently spirally stable, the friction force in the controls has usually been found to be sufficient to hold the airplane in a banked turn that, without pilot intervention, would lead to a spiral dive. A more sophisticated system, consisting of an autopilot with sufficient force output to readily overcome control system friction, is required to make an airplane spirally stable in practice. This example is typical of studies that I made, occupying less than a week of work, that revealed information of value over the years in discussions with manufacturers or with other workers at Langley who were not always familiar with the practical aspects of airplane operation.

Snaking Oscillations

Another stability problem that was quite common in airplanes of the period around WW II was a tendency for a continuous small-amplitude lateral oscillation in straight and level flight. This problem was called "snaking" and its cause was quite mysterious. Among the explanations offered were response of the normal lateral oscillation of the airplane to continuous small-amplitude turbulence, periodic flow separation from the wing root that affected the vertical tail, or nonlinear aerodynamic characteristics of the wing or tail surfaces for small changes in angle of attack. One explanation, which will be discussed subsequently, was the unsteady lift characteristics of the vertical tail at low frequencies.

While some of these explanations may have had some influence in rare instances, the true explanation was first given by George Schairer of the Boeing Company in an analysis of this problem on the Boeing 314 flying boat, one of the China Clippers. He pointed out that at small angles of sideslip, the rudder had a tendency to float against the relative wind, which caused the airplane to swing around and yaw in the opposite direction. Friction in the rudder system, however, held the rudder in this position as the airplane swung through zero sideslip. On reaching a sideslip in the opposite direction, the rudder hinge moments would eventually break through the friction force and the cycle would be repeated in the opposite direction. Thus, energy was fed into the oscillation by the rudder, which caused the oscillation to build up to an amplitude where this energy equaled that removed by the inherent damping of the airplane.

On learning of this explanation, efforts were made to verify it. A convenient test airplane was the Fairchild 22 on which an experimental all-moveable vertical tail had been installed. This type of tail surface was an invention of Robert T. Jones and had the advantage that hinge moments due to angle of attack and due to deflection could be adjusted separately with changes in the hinge location and tab gearing. The tests were made covering a range of conditions and friction values, and the validity of the theory was established (ref. 4.8).

The question arises as to how such an apparently obvious control motion could have escaped detection. The explanation is that because of the relatively low damping of the Dutch roll oscillation, the rudder motion required to sustain a constant-amplitude oscillation is only a small fraction of the amplitude of the yaw or sideslip. For example, in a typical snaking oscillation of plus or minus two degrees of sideslip, the rudder motion required might have been only plus or minus two-tenths of a degree. This small motion was less than the sensitivity of control position recorders used at that time, and this motion could be absorbed by stretch in the control cables without being felt at the pilot's rudder pedals.

Another little-known aspect was the tendency of the rudder to float against the relative wind at small sideslip angles. Most control surfaces float with the relative wind at larger sideslip angles. In typical wind-tunnel tests, measurements had been made at increments of angles of attack or sideslip of five degrees, and as a result, the small changes in characteristics at very small values of angle of sideslip were not detected.

In addition to flight tests, theoretical studies were made to explain and quantitatively predict the oscillation. These studies are discussed in a subsequent chapter.

Rudder Deflection Required for Trim

For many years, aeronautical engineers instinctively built airplanes with a symmetrical configuration. The aerodynamic symmetry is destroyed, however, by the engine torque and the rotation of the slipstream. On many early airplanes, the fin was offset to attempt to align it with the rotating slipstream and to reduce the rudder forces to maintain trim in takeoff and climb conditions. With the greatly increased power and speed range of fighter airplanes in WW II, the ability of the rudder to trim the airplane became a serious problem. If the fin was offset to attempt to trim the airplane at low speed, the rudder pedal force to maintain an aim point on a target in a high-speed dive frequently exceeded 100 or even 200 pounds. The pilots could exert these forces, but the aiming accuracy was found to be very adversely affected by the large and varying forces.

Model airplane builders had long ago encountered the problem of very high propeller torque on rubber-powered models. The effect of this torque relative to the aerodynamic forces on the wings was larger than

that encountered on even the highest powered fighters because the models had power to climb vertically, whereas the fighters did not. The model builders, less inhibited by their training than full-scale airplane designers, had found that the propeller torque could be handled by offsetting the wing laterally. I therefore did not hesitate to suggest this solution for the problem on the full-scale aircraft. Tests were run on the Brewster XSBA-1, which provided another useful research application of this airplane.

Though the wing could not be offset laterally, the same effect could be obtained by shifting the center of gravity laterally by unsymmetrical loading of the wing fuel tanks. For a conventional propeller turning clockwise when viewed from the rear, the center of gravity should be offset to the right. The method was found to reduce greatly the rudder deflection for trim in low-speed, high-power conditions (ref. 4.9). This method is highly effective because three separate effects combine to reduce the rudder deflection. The reduced aileron deflection reduces the adverse yaw, the gravity component along the longitudinal axis provides a yawing moment, and the left sideslip required to maintain straight flight is reduced. (This sideslip is required to offset a combination of the side force on the fuselage due to slipstream rotation and the side force on the vertical tail due to rudder deflection.) All these effects are in the direction to reduce the rudder deflection. In high-speed dives, use of an offset fin or any other structural deflection gives a moment that increases as the square of the airspeed, which results in rapidly increasing rudder forces to maintain trim, whereas the offset center of gravity provides a moment that does not vary with speed. These effects had been known to some airplane designers as early as WW I, but the effectiveness of this technique on high-powered airplanes was not widely recognized.

Progress in Flying Qualities Research

My work on flying qualities research continued during and after the period of WW II. As head of the Stability and Control Section, I had the responsibility for analyzing the data and writing reports on many of these tests. I did not do this alone, of course. By the time I had worked at Langley for 3 or 4 years, the measurement of flying qualities had become a somewhat routine procedure, and the members of the section were assigned as project engineers on the various airplanes under test. The results of these studies were still of great interest, because something new about handling qualities was learned from nearly every airplane that was tested. A great deal of my time, however, was spent in reviewing and correcting the various reports that were produced and in bringing their quality up to the high standards that had become traditionally associated with NACA reports.

One aspect of work at NACA that I considered very desirable was that there were always many challenging technical problems to work on, but if at times I lacked the inspiration or ambition to tackle such problems, there was always plenty of routine work to occupy the time. Reviewing the reports put out by the engineers was one such task. Many of the engineers hired during the period of rapid expansion during WW II lacked the extensive technical background or report-writing ability to meet the standards established in earlier years. I spent a lot of time with the engineers with less training to improve their skills. At the time, this seemed a thankless task, but in later years many of these engineers went on to high positions in industry or in the space program, and I was credited by some of the engineers with helping them to gain some essential knowledge required for their careers.

Meanwhile, I worked on many specialized technical problems. One such activity was relating the design features of the airplane in a rational way to the measured handling qualities. Gilruth's research had allowed many of the handling-qualities requirements to be stated in quantitative terms, such as the control-stick motions or forces required for specified maneuvers and flight conditions. I developed theories to predict these quantities based on the dimensions and design features of the airplane. As an example, I calculated the variation of control force and control position with airspeed in straight flight or with normal acceleration in maneuvers. Included in these calculations were the effects of such features as springs or mass balance weights often incorporated in control systems. A flight study was made in the XSBA-1 airplane to verify these calculations. This analysis required nothing more than algebra, and though some engineers were no doubt familiar with these effects, it is remarkable that little emphasis was given to these relations in aeronautical engineering courses. Before Gilruth's research, however, the incentive for making such analyses did not exist.

Starting in 1943, I taught evening courses to newer Langley employees on flying qualities and on stability and control of airplanes. These courses were part of the University of Virginia Extension Program that allowed new employees to study for advanced degrees. The notes for the flying qualities course were later published as a NASA Report entitled *Appreciation and Prediction of Flying Qualities*, a name suggested by Hartley Soulé (ref. 4.10). This report was later used by many other organizations such as the Navy Test Pilot School at Patuxent River, Maryland, and by colleges, including my alma mater, MIT. Gilruth's report, *Requirements for Satisfactory Flying Qualities*, had become the basis for flying qualities requirements of the Air Force and Navy. The military services incorporated in their requirements many findings of the NACA studies of flying qualities, in addition to some more stringen provisions based on maneuvers required in air combat and other military operations. These requirements

FIGURE 4.8. Grumman XTBF-3 airplane, a torpedo bomber used in the war in the Pacific. This airplane had near neutral directional stability at small sideslip angles (0 to 5 degrees) caused by blanketing of the vertical tail by the disturbed flow behind the large greenhouse canopy. The use of a ventral fin restored the directional stability at larger sideslip angles so that the characteristics were not considered dangerous.

were revised from time to time and I, along with other Langley researchers, was required to confer with officials of the Air Force Flight Dynamics Laboratory and the Navy Bureau of Aeronautics to discuss the revisions to the requirements and to write comments on the changes.

Static Longitudinal and Directional Stability

One of the more important considerations in designing a new airplane is selecting the size and configuration of the horizontal and vertical tails. Gilruth's requirements for satisfactory flying qualities had given a rational basis for the design of these surfaces, but the application of these requirement involved the calculation of aerodynamic forces on all parts of the airplane and the interference effects of the airplane on the tail surfaces.

Before I started work at Langley, Gilruth and Maurice D. White had written a report in which the longitudinal stability was predicted on the airplanes that had been used in the handling-qualities measurements (ref. 4.11). The quantity used to compare the predicted and measured longitudinal stability was the variation of elevator angle with angle

of attack. This quantity was used because it is measured in flight, whereas the variation of pitching moment with angle of attack, which could be obtained in a wind tunnel, was not available in the flight tests. Remarkably good agreement was obtained between the predicted and flight measured values. This agreement resulted from careful inclusion of all sources of pitching moment, such as the forces on the propeller, the influence of wing upwash on the propeller and fuselage, and the flow field at the tail. The flow field at the tail was based on the detailed studies made in the NACA Langley Full-Scale Tunnel by Katzoff and Silverstein, an excellent example of basic research with applications to airplanes long after the research was conducted. In later years, many other reports were written on prediction of longitudinal stability, but none provided any improvement in the accuracy obtained by Gilruth and White.

The design of the vertical tail appeared to be an even more critical problem than the design of the horizontal tail. Many airplanes tested in flight had proved to have deficient directional stability and as a result failed to meet the flying qualities requirements (figure 4.8). As a result, I attempted to make an analysis of directional stability similar to that made by Gilruth and White for longitudinal

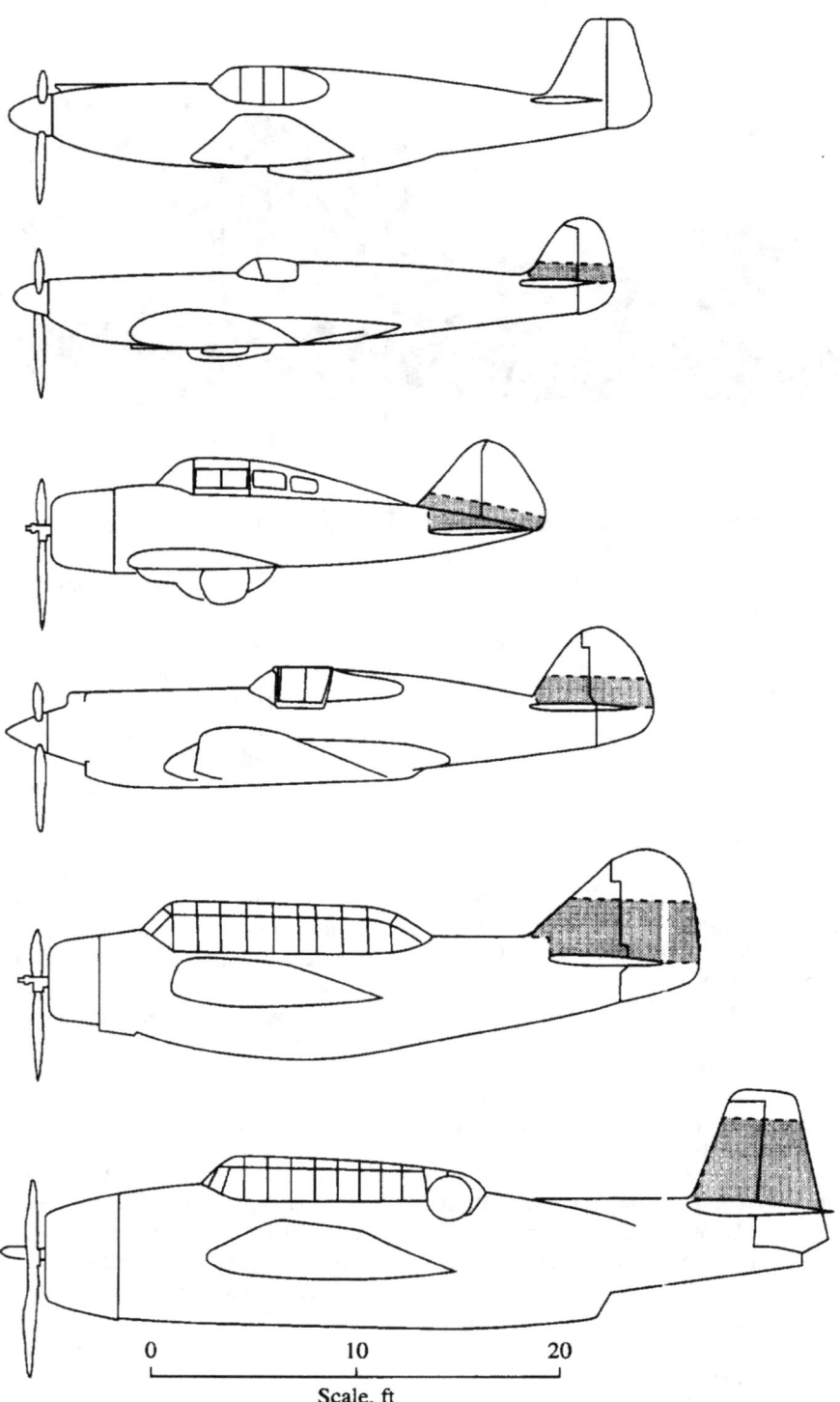

FIGURE 4.9. Portion of tail area that must be considered ineffective to obtain agreement between calculated and measured variation of rudder angle with sideslip angle (low-wing airplanes).

0   10   20

Scale, ft

stability. The quantity used for the comparison with flight data was the variation of rudder angle with sideslip in steady sideslips, a quantity analogous to the variation of elevator angle with angle of attack in the longitudinal case. This analysis was written as a proposed report and was reviewed by members of the Stability and Control Division. Their conclusion was that the report should not be published because the analysis was too arbitrary and was not based on sound theory or data.

An example of the arbitrary nature of the analysis is given in figure 4.9, taken from the report, which shows the amount of vertical tail area that must be considered blanketed by the flow over the fuselage and canopy to give agreement with the flight tests. After examining this figure (and a similar one for high-wing airplanes), the lines defining the part of the tail considered ineffective were drawn according to a set of rules that considered the height of the canopy and whether it had sharp or rounded corners. Obviously, the airplanes with large "greenhouse" canopies lost a lot of directional stability because of the poor flow over the vertical tail. Unfortunately, wind-tunnel data for the effects of the fuselage on the flow at the vertical tail were not available. As a result, accurate analysis of this effect could not be made. Despite the arbitrary nature of the rules defining the effective vertical tail area, the analysis succeeded quite well in predicting the measured

variation of rudder angle with sideslip on most of the 19 airplanes that were used in the comparison.

I have often felt that the report should have been published, despite its use of arbitrary methods, because the report included many factors that had previously been neglected in designing vertical tails. The report would also have encouraged wind-tunnel researchers to investigate systematically some of the configuration features that led to the arbitrary rules, thereby allowing a more rational analysis. No similar report based on wind-tunnel data was ever published. Fortunately, the designers soon found the advantages of much cleaner canopies, such as bubble canopies, in reducing drag so that future airplanes did not have such poor flow conditions at the vertical tail. In recent years, methods based on computational fluid dynamics allow accurate calculation of the flow field over the entire airplane. These methods are known as panel methods, as the entire surface of the airplane is represented by a large number of flat panels that approximate the true surface contours. These methods neglect flow separation, but result in more accurate predictions of longitudinal and directional stability because airplanes have become cleaner as a result of refined aerodynamic design and the use of jet engines. The need for an analysis similar to the one I attempted has therefore decreased.

CHAPTER 5

Analytical Studies

As in most scientific and engineering work, analytical studies are required in conjunction with experiments to understand the results of aeronautical research and to predict the characteristics of new aircraft or aeronautical systems. Much of the progress in improving the performance and safety of airplanes relied heavily on analytical work, which in turn depended on the availability and understanding of analytical techniques. This section outlines the status of analytical techniques that I had encountered in my studies and that were generally available to aeronautical researchers at the time I started employment with the NACA in 1940.

Studies of the stability and control of airplanes, my field of specialization, relied almost completely on the application of Sir Isaac Newton's laws of dynamics, which in turn led to the need to solve differential equations. In my college courses in mathematics and physics, Newton's laws of motion were among the first subjects studied, and the various means to apply them in different scientific disciplines occupied most of the subsequent curriculum. Mathematics courses, particularly differential and integral calculus, were presented to students as a general preparation for all the courses. These methods were originally developed largely by Newton to solve his own problems. The present notation for differential and integral calculus was originated by Jacob Liebnitz, a contemporary of Newton. A branch of calculus called differential equations occupied a whole term at MIT. Many of the methods that were taught to solve differential equations, however, applied to special forms of equations that did not occur in connection with airplane stability and control. This subject requires the solution of simultaneous linear differential equations with constant coefficients. The standard mathematics curriculum did not give this subject any special emphasis. As a result, it was left for the professors in the aeronautical courses to emphasize the importance of this particular equation. These professors, in my courses at MIT, were practical engineers without a strong mathematical background. As a result, graduates were provided with only a minimal introduction to the branch of mathematics most useful for their subsequent work. I feel that this situation existed in most of the colleges in the United States at that time, with the result that engineers who wanted to go into this field had to spend a great deal of their time in reviewing work done largely by electrical engineers or by some aeronautical researchers in Europe. Many chose to put their emphasis on other areas with which they were more familiar. The result was undoubtedly a slowing down of progress in this field.

The reason that simultaneous differential equations with constant coefficients arose in aeronautical stability work may be explained in somewhat more detail as follows. For each variable that describes motion (the degrees of freedom), a differential equation is set up. In the case of longitudinal motion of a rigid airplane, for example, the variables would be vertical displacement, horizontal displacement, and pitch angle. Newton's third law states that force equals mass times acceleration. Thus for each variable, the aerodynamic or gravity forces due to the motion, which may produce forces or moments proportional to the displacement or velocity, are equated to the mass times the acceleration of that variable. Acceleration is represented by the second derivative of the variable with respect to time, velocity the first derivative, and displacement does not involve a derivative. As a result, a differential equation of the second order is written for each degree of freedom, which results in three simultaneous differential equations. These equations, in the usual formulation, are linear for the following reasons.

The airplane is first considered as flying in a trimmed condition. After a disturbance, the forces on the airplane change as a result of the effects of changes in angle of attack or airspeed on the various components of the airplane. These changes are necessarily small in the normal unstalled range of flight. The stall angle, at least on airplanes of WW II vintage or earlier, was usually about 15 degrees, and most disturbances producing forces within the structural capability of the airplane would be much less than this value. Ideal fluid theory shows that within this range, forces vary nearly linearly with angle of attack. The only factor that would change this condition would be the effect of the boundary layer, but the boundary layer on full-scale airplanes designed for efficient flight is so thin that it causes little effect on the forces. Likewise any change in airspeed caused by a disturbance is likely to be small compared to the initial large value of airspeed. Therefore, though the forces vary as the square of the airspeed, the variation in force for a small change in airspeed may be

considered to vary almost linearly with the change in airspeed. By comparison with many mechanical systems, the airplane as a whole does not have any coulomb friction tending to hold it in its trimmed condition. Coulomb friction is the force caused by rubbing two solid surfaces over each other, and from an analytical standpoint, is usually considered to be a constant force that is independent of velocity and opposing the motion. The force caused by aerodynamic effects is in marked contrast to that appearing in mechanical systems. This difference results in the validity of linear equations to describe the motion of an airplane, whereas linear equations are usually a poor representation of mechanical systems.

Solution of the system of linear differential equations with constant coefficients can be accomplished by the classical method given in most books on differential equations. These solutions go back to the work of mathematicians in the earliest days of mathematics and were perhaps first summarized in a paper by Leonhard Euler in 1739. This solution is considered elementary from a mathematical standpoint, inasmuch as it involves elementary functions such as exponentials and sine and cosine functions. From the standpoint of the practical engineer using the equipment available when I came to work at Langley, which consisted of slide rules or mechanical calculators, the solution is very time-consuming. Determining the characteristics of the various modes of motion, such as the phugoid oscillation or short-period mode, requires solving for the roots of a fourth-degree algebraic equation. If the effects of a simple autopilot are included, a sixth-degree equation results. This problem can be solved only by trial and error or by methods of successive approximations. Further lengthy calculations are required to determine the constants giving the amplitudes of each mode for known initial conditions.

Aeronautical engineers confronted with these problems made efforts to devise simpler methods for solving for the roots of equations, as will be described in the subsequent material. At the same time, and usually without the knowledge of aeronautical engineers, electrical engineers had developed different approaches that allowed practical solutions of even more complex systems of linear differential equations. Such solutions were required in the design of electrical circuits, vacuum-tube feedback amplifiers, and transmission lines. One method, first summarized in a book by Hendrik W. Bode of the Bell Telephone Laboratories, entitled Network Analysis and Feedback Amplifier Design (ref. 5.1) exploited the use of sinusoidal forcing functions of various frequencies to the dynamic systems under consideration, from which the stability of the systems could be determined. This method was slow to be discovered by the aeronautical engineering profession because of the unfamiliar notation and applications of the electrical engineers. Later, the method was called the frequency-response method and was widely used. A second approach with the general title operational methods was also introduced by the electrical engineers. A British electrical engineer and mathematician named Oliver Heaviside devised a system of operational calculus about 1887. The Heaviside operational calculus was publicized in this country in a book by Vannevar Bush, entitled Operational Circuit Analysis (ref. 5.2) and was introduced to the aeronautical profession by Robert T. Jones in NACA report No. 560, A Simplified Application of the Method of Operators to the Calculation of Disturbed Motions of an Airplane (ref. 5.3). Later, the book by Murray F. Gardner and John L. Barnes, Transients in Linear Systems Studied by the Laplace Transformation (ref. 5.4), described the operational method based on Laplace transforms, which became generally accepted as easier to understand than the Heaviside method. The advantage of operational methods is that solutions for frequently encountered equations and for special inputs such as steps and ramps can be obtained from tabulated precalculated formulas.

A third approach for solving these systems of equations was the use of simulators, generally referred to at the time of their development as differential analyzers. Vannevar Bush developed a mechanical differential analyzer at MIT that I saw about 1934 while I was still in high school. This machine took the equations in an integrated form so that the various terms required integration rather than differentiation. The integration was done by devices known as rolling-wheel integrators. The low torque output of these devices was amplified by winch-type electromechanical servos to drive a large array of shafting and gearing that allowed setting in the correct constants from the equations. The whole equipment required a room about 25 by 60 feet. This machine had excellent accuracy and was used by an MIT student to get some of the first solutions for the motion of airplanes with automatic controls. Later, such simulators were made with servo-driven potentiometers to enter the constants and electronic operational amplifiers to perform the integrations. These machines had tremendously increased capabilities, but they in turn became obsolete with the development of electronic digital computers.

Although the airplane as a whole can be described accurately by linear differential equations, many of the subsystems, such as control surfaces or autopilots, involve nonlinear components. For example, control surfaces have static friction, and electronic devices often have switches that give a discontinuous output. In general, nonlinear systems include devices in which the output varies in a nonlinear but continuous manner, such as a crank, or in which the output varies discontinuously, such as a bearing with coulomb friction or an on-off switch. Much of the mathematical analysis of nonlinear systems when I was in college had been concerned with the continuous type of nonlinear systems, though solutions were usually available only for systems capable of representation by special types of differential equations. The only method of analysis for discontinuous systems with which I was *familiar was the phase plane method, a graphical method described in the book by Nicolai Minorsky,* Introduction to Non-Linear Mechanics *(ref. 5.5). I used this method on problems involving simple types of autopilots. With the eventual development of analog computers, such problems could be solved readily.*

A final point that should be appreciated by the reader is the state of computational facilities at the time much of my work at Langley was conducted. Prior to about 1955, the only widely used computers were the slide rule and mechanical calculators such as the Marchant and Frieden that required several seconds to multiply or divide two numbers. To perform lengthy calculations, female employees called computers were employed to calculate results with these machines by following sheets that had the necessary steps listed in tabular form. Such calculation sheets are today called spreadsheets.

In making analytical studies of problems, a very desirable result was a closed-form solution, which means a formula in which numbers can be substituted for any particular case to determine the numerical value of the desired quantity. The determination of these closed-form solutions had been the objective of mathematicians and scientists for many years. In the aeronautical field, for example, Ludwig Prandtl determined the formulas for calculating induced drag of wings, Max Munk derived formulas for moments acting on ellipsoids in steady flow, and Theodore Theodorsen derived formulas for the lift and moments on oscillating wings. The formulas are considered to be in closed form if they give results in terms of known tabulated functions, such as trigonometric functions or Bessel functions. Most of these formulas and their originators became very famous because airplane designers could calculate results accurately for a configuration that at least approximated the one in which they were interested.

With the advent of analog computers and later of high-speed digital computers, the need for closed-form solutions was reduced. Most calculations on the digital machines are made by numerical methods. One such method is the step-by-step method in which the response of a system to a disturbance or the trajectory of a vehicle is calculated a small increment at a time. Another numerical method is the Monte Carlo technique, in which many trial solutions are made to find the one with the best answer for the problem. Many problems of structures or fluid mechanics are solved by finite-element methods, in which the equations relating each small element of a large structure or flow field are solved. These methods involve tremendous amounts of numerical computation, which modern electronic computers can handle in a very short time. Since I had none of these methods in my college education, my facility with such techniques is much less than that of students who grew up in the computer age.

With this background I will describe some of the analytical studies that I conducted during my early employment at Langley.

Graphical Solution of the Quartic Equation

As pointed out in the introductory section, the solution for the motion of an airplane following a disturbance was very tedious because of the numerical calculations required. One of the problems encountered was solving for the roots of higher degree equations. A fourth-degree equation resulted in the solution for the longitudinal motion of a rigid airplane. For many problems, this equation could be reduced to a second-degree, or quadratic, equation by considering the airspeed constant, a valid assumption if short-period maneuvers were being considered. The lateral equations likewise result in a fourth-degree equation if certain simplifying assumptions are made. These fourth-degree equations, also known as quartics, are called the characteristic equations for the systems, and solving for the roots of these equations is the first step in calculating the response of an airplane to controls.

Quadratic, cubic, and quartic equations may be written as follows:

quadratic $\quad ax^2 + bx + c = 0$

cubic $\qquad ax^3 + bx^2 + cx + d = 0$

quartic $\quad\ \, ax^4 + bx^3 + cx^2 + dx + e = 0$

High school students of mathematics are familiar with the formula for the solution of a quadratic equation. Formulas also exist for the solution of cubic and quartic equations, though they are considerably more complicated than those for a quadratic. For this reason, graphical methods or methods of successive approximations have been sought for the solution of these equations. In a paper on propeller governors, Herbert K. Weiss presented a set of graphs for the solution of cubic equations in terms of two parameters calculated from the coefficients of the equation (ref. 5.6). Shih-Nge Lin, in appendix I of his MIT thesis, described a method that he developed for a solution of quartic and other higher order equations (ref. 5.7). This procedure is a method of successive approximations that uses repeated long division. This method, though it was considerably quicker than the classical techniques, was still quite time-consuming without considerable practice. I therefore attempted to develop a set of charts for the quartic similar to those of Weiss for the cubic. A brief summary of this method and a sample copy of the charts are presented as appendix IV. These charts were never published. I used them to some extent in my work, but found that the time to solve for the parameters used in the charts and then to return the solution to an expression involving the original variables, required about as much time as Lin's method. This method is presented mainly because it was my only excursion into pure mathematics. With the development of high-speed computers, of course, the engineer no longer has to be concerned with these calculations because readily available computer programs can solve quartic or even much higher degree

equations in a fraction of a second. As late as 1962, however, there was still interest in simplified methods for determining the roots of algebraic equations, as shown by a report by James W. Moore and Rufus Oldenburger of which I have an unpublished copy. This report presents a systematized procedure similar to Lin's method and analyses problems of convergence for cases, such as unstable roots, in which convergence of the method may be slow. Oldenburger was a well-known expert on servomechanisms before the subject of automatic control became a favorite subject for control theorists.

The airplane with fixed controls in the unstalled flight regime, fortunately, is beautifully linear. That is, all the aerodynamic forces and moments increase in proportion to the magnitude of the displacement or angular velocity, even down to very small magnitudes of motion. As a result, the measured motion of airplanes had been found to be closely predicted by the theory. Another favorable feature of linear systems is that a given solution is applicable to all magnitudes of motion. An increased initial disturbance simply increases all quantities involved in proportion to the disturbance without changing their time dependence. This condition no longer exists with nonlinear systems, and a separate solution has to be obtained for each magnitude of motion.

Solution of a Nonlinear Problem

The problem of snaking oscillations, discussed in the previous chapter, is an example of a nonlinear problem. The nonlinearity arises because of friction in the rudder control system. The rudder, instead of moving in proportion to the motion of the airplane, sticks until the aerodynamic hinge moments exceed the friction. Then the rudder starts to move under the influence of both the aerodynamic forces and the friction force.

By the time the snaking motion had been explained, I was familiar with another method to analyze nonlinear systems. Robert T. Jones, while analyzing the stability of some of the first guided missiles developed by the Navy, had devised a technique assuming that the vehicle performs a sinusoidal oscillation. The actual control motion resulting from this sinusoidal motion was then calculated or measured experimentally. He then calculated the response of the system under the assumption that the work done on the vehicle per cycle must be the same as that done by the action of the control, and the angular impulse imparted to the vehicle over a half cycle must equal the change in angular momentum of the vehicle caused by the operation of the control. The work and momentum relations give two equations from which the frequency and amplitude of a constant-amplitude oscillation may be calculated. The expressions for the work and momentum imparted by the control may be shown to be related to the lowest order cosine and sine components of the motion of the control when expressed as a Fourier series. Jones' technique was therefore equivalent to the frequency-response method, which was then generally unknown to aeronautical engineers. It was later found that it had been developed to an extensive degree with different notation by electrical engineers for studying the stability of feedback amplifiers. I was intrigued by this method, because it appeared that the exact form of the control surface motion when subject to static friction would be important. As a result, I attempted to analyze the snaking problem by this method. I first tackled the problem of aileron snaking, a steady oscillation in roll caused by friction in the aileron control system. This problem had never been encountered in practice because of the much larger damping of the roll subsidence mode of an airplane when compared with the damping of the lateral oscillation. A much larger tendency for the controls to float against the relative wind would have been required to produce an aileron oscillation than to produce a rudder oscillation. The problem was easier to

formulate, however, because of the simpler equations governing the rolling motion. Even in this case, calculation of the aileron motions subject to the control friction and the aerodynamic moments on the aileron proved to be quite difficult. After working on the problem as a part time activity for several months, I succeeded in obtaining a closed-form solution. A typical example of possible steady oscillations for two cases is given in figure 5.1. In the first case, the friction is enough to cause the aileron to stick during part of the cycle; in the second case, the aileron motion is continuous.

After going through all this work, I concluded that it was hardly worthwhile to attempt a similar solution for the more practical case of the rudder snaking oscillation. The results on the aileron oscillations were never published. Many years later, however, a problem was encountered of control of a missile in which the ailerons were operated by a gyroscope sensing rolling velocity. This problem was exactly equivalent to the problem that I had solved. The only difference was that the hinge moments proportional to rolling velocity were applied to the ailerons by the gyroscope rather than by an aerodynamic floating tendency. It was possible to use my analysis to obtain some design data for the missile. As a result, my analysis was not completely wasted after all.

The frequency-response technique is based on the assumption that although the control motion is nonsinusoidal, the response of the controlled vehicle is very nearly sinusoidal because the relatively large inertia and slow response of the vehicle prevents it from responding appreciably to the higher frequencies in the irregular control motion. This assumption justifies the calculation of the control motion that is based on a sinusoidal vehicle motion. In most discussions of this method, however, the use of the lowest harmonic of the control motion in calculating the vehicle response is not justified on physical grounds. Jones' independent approach, which first based the response of the vehicle on the momentum and work relations over a

FIGURE 5.1. Time histories of aileron snaking oscillations as calculated by my theory.

(a) Discontinuous aileron motion (top).

(b) Continuous aileron motion (bottom).

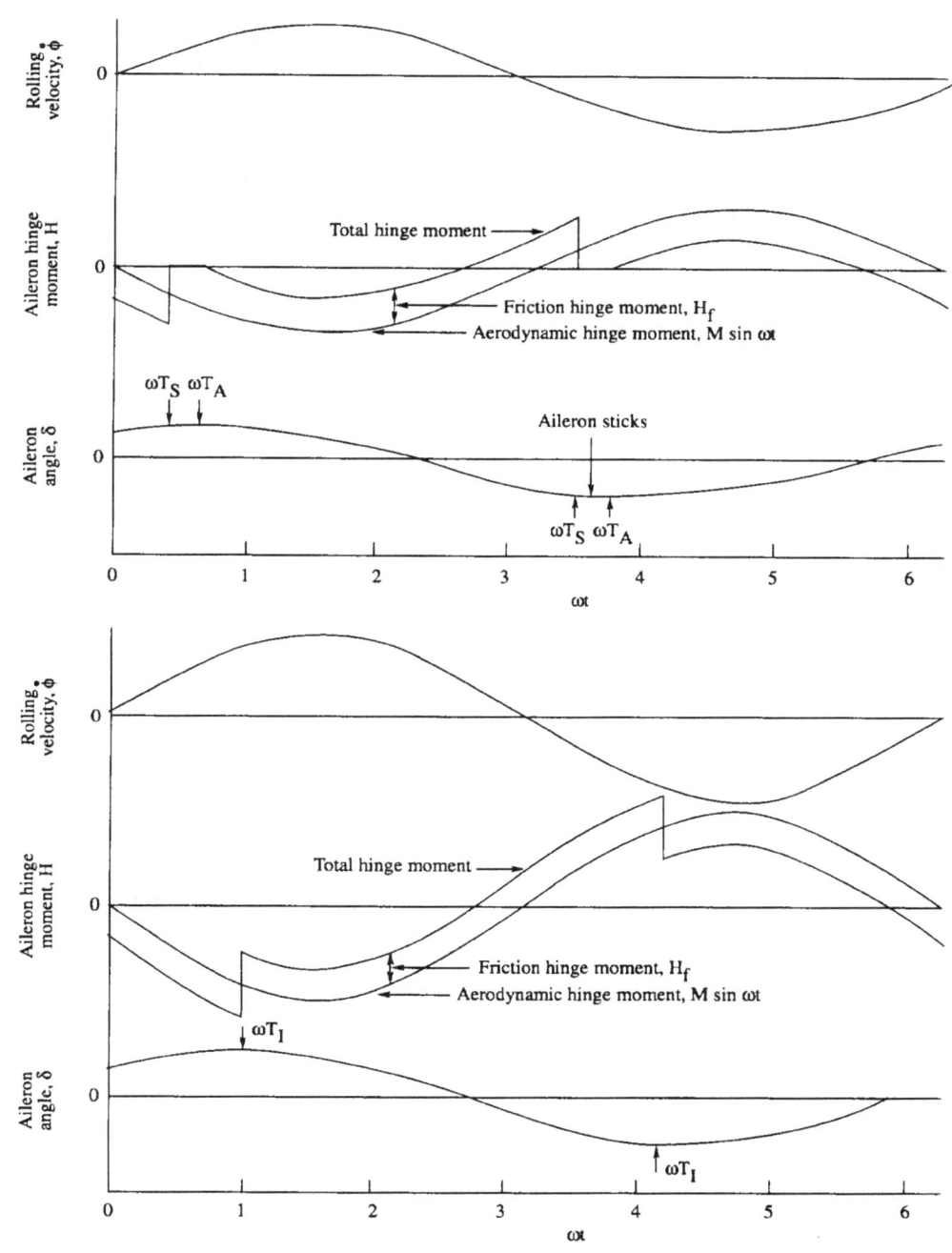

cycle and later showed that these relations gave expressions corresponding to the lowest order harmonic of the control motion, gives a clear physical interpretation of the application of the frequency-response method to nonlinear systems and provides insight into the accuracy of the method for this purpose. In the ensuing years, methods based on the frequency-response method were extensively used at the NACA and elsewhere for design of missile control systems, determination of stability derivatives from flight data (now called parameter identification), and calculation of response to turbulence. The many developments of the method, such as Bode plots, the Nyquist criterion, and root locus techniques gradually became part of the aeronautical engineer's mathematical equipment.

Computation of Lateral Oscillation Characteristics

As pointed out previously, the calculation of the airplane modes of motion prior to the introduction of high-speed computers was a tedious process. The solution for the lateral modes of motion requires the solution of a quartic equation. To simplify the process, approximate procedures were developed. Simple expressions could be found for the characteristics of the spiral mode and the roll subsidence, but accurate calculation of the Dutch roll usually required solution of the complete equations. A graphical method, first developed in 1937 by R. K. Mueller of MIT, proved to be applicable to the Dutch roll mode. Mueller first used the method in developing the world's first electric analog computer, which he used for the calculation of the longitudinal motion of airplanes (ref. 5.8). He discovered, however, that once he had perfected the graphical method, he no longer needed the analog computer. Later the method was discovered independently by K. H. Doetsch and W. J. G. Pinsker, two former German engineers working at the

Royal Aircraft Establishment (RAE) after WW II. They called the technique the time vector method and applied it to calculation of the Dutch roll mode. Later, W. O. Breuhaus of the Cornell Aeronautical Laboratory, on a visit to England, learned the method from the RAE engineers and publicized it in the United States (ref. 5.9). Prof. E. E. Larrabee of MIT, on a sabbatical leave at the NACA Langley laboratory, showed how the method could be used to calculate stability derivatives from measured lateral oscillation data (ref. 5.10).

The time vector method is based on the fact that in a free oscillation of a linear system, the variables involved always maintain the same ratios of amplitude and the same relative phase angles. The variables can be depicted on a polar vector diagram in which the amplitude of each variable is shown by the length of an arrow and the relative phase angles by the directions of the arrows. In a damped oscillation, the diagram would rotate and shrink, but always maintain the same relative magnitude and angular separation of the vectors. All information about the motion could therefore be obtained by taking a snapshot of the diagram, which showed it at a given instant of time. Fixed relations can also be calculated between displacement, velocity, and acceleration of any quantity based on the frequency and damping of the motion.

The equations of motion involve a series of terms in each equation. In the lateral case, the equations involve roll, yaw, and sideslip. Each term in the equations consists of one of these variables multiplied by a stability derivative which gives the variation of the force or moment with the variable. For example, a typical term might be rolling velocity multiplied by the derivative, variation of rolling moment with rolling velocity. All the rolling moments contributed by the different variables are added up to give the total rolling moment, which must equal the inertia in roll multiplied by the rolling acceleration. This term may be placed on the opposite side of the equation, which results in a sum of terms equal to zero.

Since the derivatives are constants, each term varies with time in exactly the same way as the variable in that term. A polar vector diagram may therefore be drawn that includes each term in the equation with the tail of each vector coinciding with the head of the previous one. Since the terms add up to zero, the diagram must close. A similar diagram may be drawn for each of the variables: roll, pitch, and yaw.

In using the method, the quantities to be determined are the frequency, the damping, and the ratios of the variables. Usually a value of unity is assigned to one of the variables, such as sideslip, and the ratios of roll to sideslip and yaw to sideslip are to be determined. All other quantities must be known beforehand or estimated. Usually the frequency is estimated first from the simple relation of the frequency of the airplane oscillating with a single degree of freedom in yaw. The damping is assumed to be zero. From these assumptions, initial values of the ratios of the variables may be determined by the closure of the vector diagrams. Then, with these values, more accurate values for the frequency and damping may be determined and these values used in a second attempt. The method is therefore an iterative procedure. Usually the convergence is very rapid and requires only two or three iterations to reach a solution.

The time vector method gained considerable recognition because of the rapidity of solution and because it gave a useful physical picture of the relations between the variables in an oscillation. It occurred to me, however, that the graphical work might be avoided if a similar convergent procedure could be performed analytically. After several trials of different methods, I discovered that the equations could be solved for the ratios of the variables. Therefore, a similar iterative procedure could be set up starting with an assumed value of frequency and a value of damping of zero, to calculate values of the ratios of the variables. I discussed this method with Bernard B. Klawans, an engineer in my branch. He worked out the details of the procedure. As it turned out, with the initial values of frequency and damping, the ratio of roll to yaw could be calculated. Then with this value and the values of frequency and damping, the ratio of sideslip to yaw could be calculated. Finally, with these two ratios, a quadratic equation for the Dutch roll root could be obtained that gave new values for the frequency and damping. Mr. Klawans checked the results for a wide range of variables. The calculations could be carried out readily on a slide rule or mechanical calculator, and the solutions converged very rapidly, usually within two or three iterations. Mr. Klawans published a Technical Note on the results (ref. 5.11).

I thought this procedure was a worthwhile improvement over the graphical method, but evidently it arrived on the scene too late. In that period, about 1956, analog computers were already available, and some digital computers had been introduced. Soon programs were provided to solve the equations of motion in a matter of seconds. As a result, the approximate procedures that had occupied the efforts of stability and control engineers for many years fell into disuse.

CHAPTER 6

Problems Encountered as a Result of Wartime Developments

The period between World Wars I and II was only 23 years. At the end of WW I, the reaction of the public against military activities resulted in large cuts in expenditures for military developments. The initial work of the NACA occurred in this period, which resulted in a substantial background of basic research information for the improvement of airplanes. The airplanes themselves, however, started as fabric-covered, externally braced vehicles with relatively low power. Without the pressure of actual wartime activity, developments by the military services occurred rather slowly.

The main incentive for increase in speed occurred as a result of various air race competitions. In England, the development of the Schneider Cup seaplane racers, engaged in international competition, resulted in high-power engines and metal construction that provided the background for the Supermarine Spitfire fighter. In the United States, a small number of sport aviation enthusiasts like the Granville brothers, who, working in a small garage in Springfield, Massachusetts, produced the fastest airplanes in the world. These racers would fly almost twice as fast as the existing pursuit airplanes of the Army Air Corps. With war threatening in Europe, these racing planes provided the incentive for improved fighters and bombers that became available in small numbers at the start of WW II.

The outbreak of hostilities in Europe showed immediately the need for increased armament and other equipment that greatly increased the weight to be carried on airplanes. Prior to the war, a newly developed pursuit airplane like the Curtiss P-36 was armed with two 30-caliber machine guns. Fighter airplanes that had been tested in combat soon were equipped with four to six 50-caliber machine guns, armor plate, bulletproof windshields, and self-sealing tanks. The result was an immediate increase in wing loading and a demand for higher power. Similar trends affected the design of bomber airplanes.

The primary effect of these changes on the flying qualities of airplanes was greatly increased control forces. The hinge moment (and hence the control force) to deflect a control surface of a given geometric shape to a given deflection varies as the square of the airspeed and as the cube of the linear dimension of the surface. This relation occurs because the load on the surface varies as its area, which increases as the linear dimension squared. The hinge moment is caused by the load multiplied by its moment arm from the hinge line, which varies directly with the linear dimension.

The result of this relation is that if the speed and linear dimension were both increased by 10 percent, the hinge moment for a given deflection would be increased by 61 percent.

If the speed were doubled and the linear dimension increased by 50 percent, a reasonable change between prewar and wartime fighter airplanes, the hinge moment for a given deflection would increase by a factor of 13.5.

Fortunately, Gilruth's requirements had shown quantitatively the satisfactory levels of control forces required for different flight conditions and maneuvers. As a result, a great deal of research was conducted at the NACA and elsewhere to provide means for obtaining the desired control forces.

The Quest for Reduced Control Forces

One of the most serious problems encountered by designers of military airplanes during WW II was keeping control forces desirably light while airplanes were being made with greatly increased weight, size, and speed. Flying qualities research had shown that maximum control forces should be kept below what a pilot could conveniently exert with one hand on the control stick or wheel. For ailerons, this force was about 30 pounds on a control stick or 80 pounds on a control wheel. Increasing the mechanical advantage of the pilot's controls was impossible because of the limited size of the cockpit and the lag in deflecting a control wheel more than plus or minus 90 degrees. Studies of aerodynamic balancing devices to reduce the aerodynamic moments on control surfaces became one of the main research objectives of wind tunnels involved in stability research.

Aerodynamic balance on most airplanes designed prior to WW II was usually accomplished by locating some control surface area ahead of the hinge line. Various arrangements of these balances are shown in figure 6.1. These balances had advantages and disadvantages from both mechanical and aerodynamic standpoints. In general, balances that were permanently located in the

air stream were subject to icing that might jam the controls. Balances that broke the contour of the airfoil added drag. In addition to such practical considerations, balances had to be selected on the basis of the hinge-moment parameters such as the variations of control surface hinge moment with angle of attack and with control deflection. These parameters had fundamental effects on the flying qualities. The effect on snaking oscillations of the variation of hinge moment with angle of attack has already been mentioned. To obtain light control forces, both of these parameters had to be reduced.

Theoretically, the control forces could be reduced to zero by reducing these hinge-moment parameters to zero, but in practice this goal could not be attained. One problem was the nonlinearity of the hinge-moment variations. For example, a control surface that was properly balanced at low deflections might be overbalanced at large deflections. A second problem that limited the degree of aerodynamic balance on large and high-speed airplanes was the effect of small changes in contour due to manufacturing differences. These differences might be almost too small to detect, yet could cause quite large changes in the control forces. The Germans, in an effort to obtain very light aileron forces on the Me109 airplane, would test fly the airplane and try different sets of ailerons until one was found that would give forces in the desired range. The British, on testing the Spitfire, mentioned encountering "rogue" airplanes that had different characteristics from the standard airplanes, the reasons for which could not be detected.

As a result of these problems, a practical limit had to be set on the degree of aerodynamic balance, which was usually 25 to 30 percent of the forces produced by an unbalanced control surface. This degree of balance, however, was nowhere near what was required to provide desirable handling qualities on the largest or fastest airplanes. In some cases, forces would have to be reduced to about 2 to 4 percent of those of an unbalanced surface.

FIGURE 6.1. Types of
control surface
aerodynamic balance.

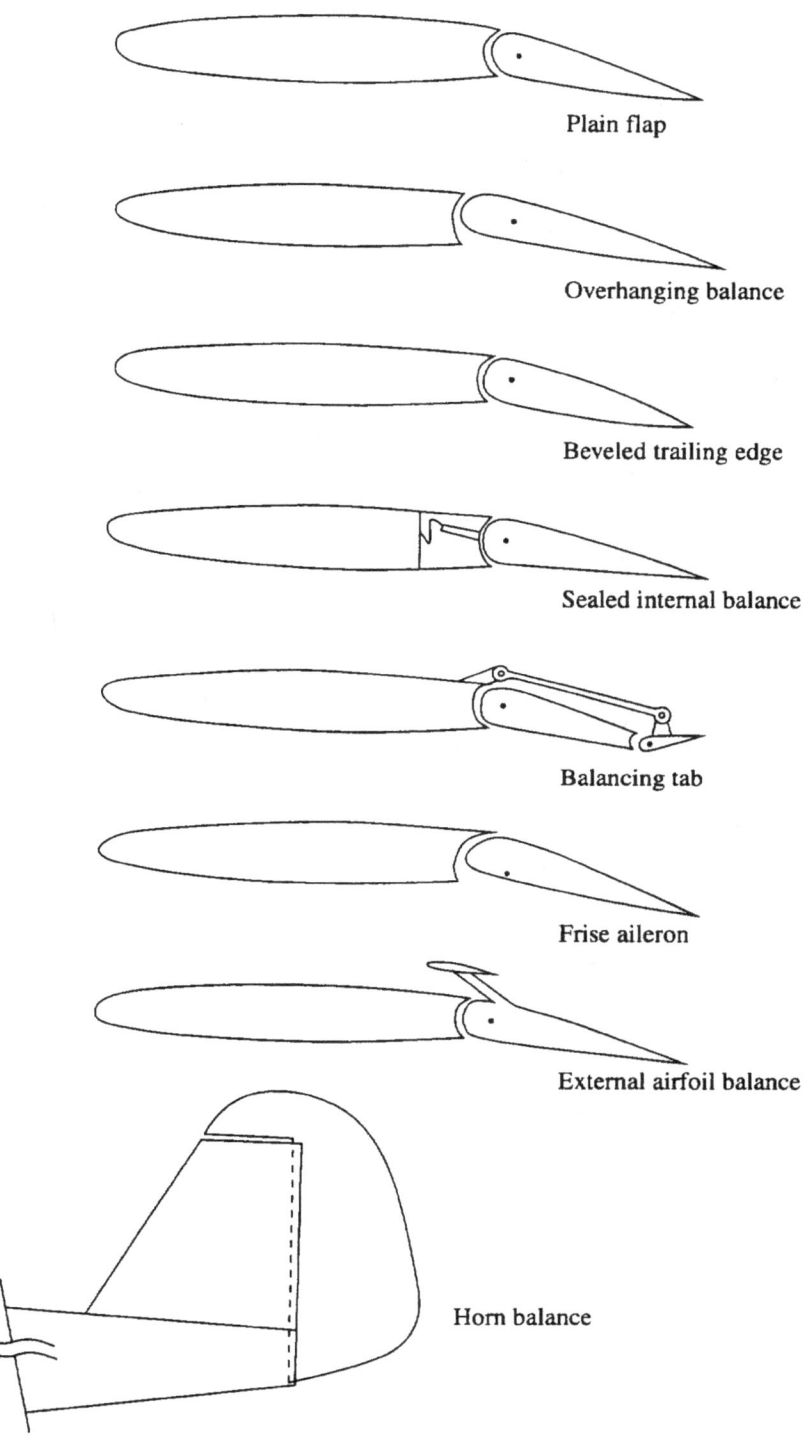

Plain flap

Overhanging balance

Beveled trailing edge

Sealed internal balance

Balancing tab

Frise aileron

External airfoil balance

Horn balance

An additional factor complicating the provision of light control forces was the universal use at the start of WW II of fabric-covered control surfaces. All control surfaces had to be mass balanced about the hinge line to avoid flutter. The fabric covering was used to reduce the weight of the surface behind the hinge line so that the mass-balance weight could also be reduced. This weight usually exceeded the weight of the control surface itself because it had to be located quite close to the hinge line. The fabric covering bulged in or out at high values of airspeed, depending on the external pressure distribution and on the way the internal volume of the control surface was vented. Usually, the fabric near the trailing edge was sucked in, which resulted in a large increase in the variation of hinge moment with deflection. To study these problems, it was necessary to know the tension in the fabric, a factor that varied widely with manufacturing techniques and with age.

Inasmuch as no instrument for measuring fabric tension was commercially available, I made an instrument that served the purpose quite well. I used a chrome-plated metal bell that was left over from the fire alarm system in the office. A glass tube was attached to the top. The bell could be sealed to the surface of the fabric with grease, and suction applied to the tube to cause the fabric to be sucked up to the shape of a spherical segment inside the bell. A small stylus rested on the fabric, and its displacement was magnified by a lever system inside the bell to move a rod up and down in the glass tube. The formula for the deflection of a spherical diaphragm under pressure then allowed the fabric tension to be calculated.

Since wind-tunnel tests had never been made in high-speed tunnels to determine the effect of contours simulating fabric deflection, some analytical work was done to determine the effect of the deflected fabric on hinge moments. Fortunately, this effort was short-lived because manufacturers soon discarded fabric covering and went to metal-covered control surfaces.

Flettner Tabs, Servo Tabs, Spring Tabs, and Whirlerons

The problems described previously with aerodynamically balanced control surfaces could be overcome if it were possible to multiply the force exerted on the controls by the pilot. Devices to provide such augmentation of force are called servomechanisms. A number of ways were devised to provide this force augmentation through aerodynamic means. Sketches of some of these aerodynamic servomechanisms are shown in figure 6.2.

An arrangement called the Flettner tab had been tried on large airplanes as far back as WW I. This device, which was invented by Anton Flettner, the same man who invented the Flettner rotor for propelling sailing ships, consisted of a small tab mounted at or behind the trailing edge of the main control surface. The pilot's control was connected just to the tab. When the tab was deflected, it moved the main control surface in the opposite direction. Because the hinge moment to deflect a control depends on the product—span times chord squared—it is apparent that very large reduction in the pilot's control effort could be obtained.

By the time I came to work at the NACA, this device, shown in figure 6.2(a), was usually called a servo tab, inasmuch as the tab acted as an amplifier to augment the pilot's force. It had the disadvantage that when the pilot moved his control stick while at rest or while taxiing, the control surface appeared floppy and did not respond as expected. To overcome this problem, a spring was placed between the control linkage and the main control surface, so that the surface would move in the desired direction even at zero airspeed. This arrangement, shown in figure 6.2(b), was called a spring tab.

Some analysis of spring tabs had been made by the British. I continued this analysis to determine how various flying quality

FIGURE 6.2. Diagramatic sketches of aerodynamically operated servo controls.

(a) Servo tab (top).

(b) Spring tab (middle).

(c) Geared spring tab (bottom).

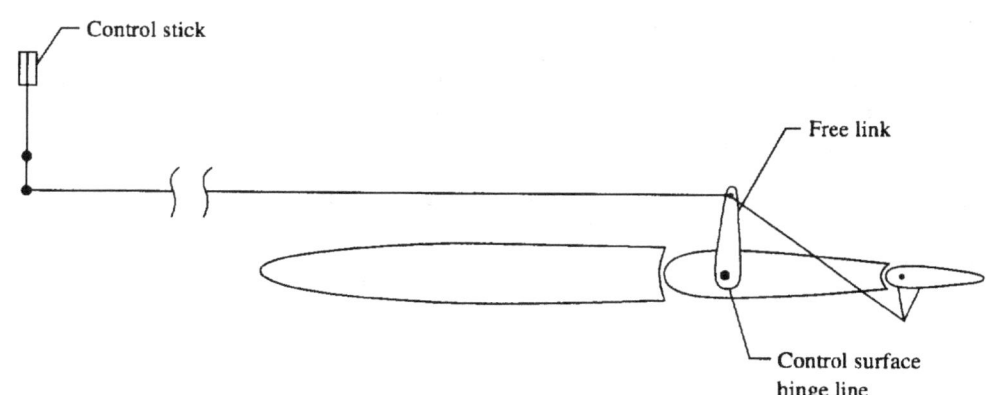

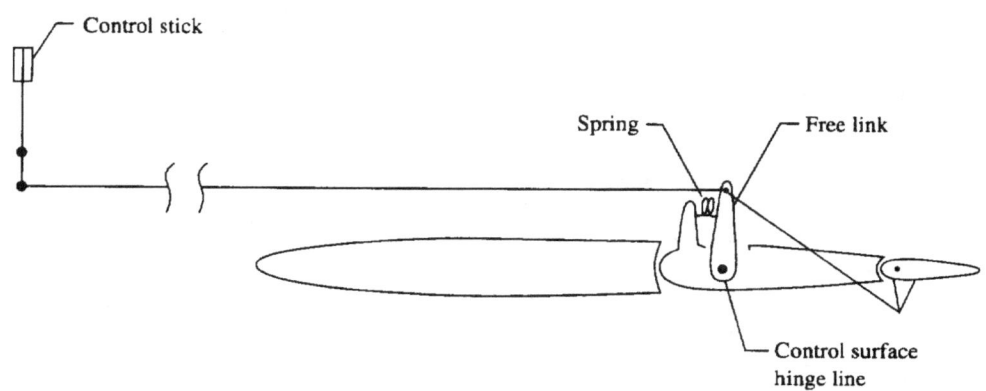

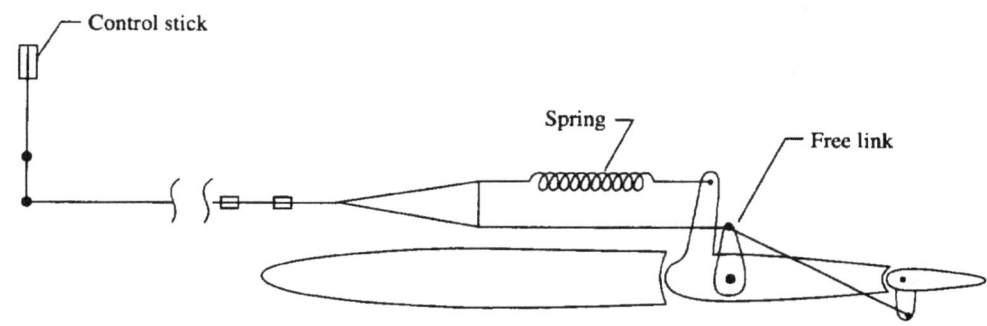

parameters would be affected by this arrangement. Because the control surface was moved by the spring at low airspeed, the variation of hinge moment with the pilot's control deflection was equal to that of an ordinary control surface. At high speed, however, most of the force to deflect the control was supplied by the tab, which resulted in a much reduced force at the pilot's control. The variation of the force reduction with speed could be regulated by the stiffness of the spring. For aileron and rudder controls, this variation of control force with airspeed was not objectionable, but it might cause some difficulties on elevator controls for the following reasons. A conventional manually operated elevator control, in which the variation of control force with deflection or angle of attack varies as the square of the airspeed, may be shown to give a value of force per g in maneuvers that is approximately independent of airspeed. The spring tab, however, gives a value of force per g that decreases with increasing airspeed. Pilots had become accustomed to a constant value of force per g. Decreased force per g at high speed might lead to a tendency to over stress the airplane. This problem may be overcome by use of the geared spring tab shown in figure 6.2(c).

A method of using a tab to reduce control forces shown previously in figure 6.1 was called a geared tab or balancing tab. That is, the tab was moved by a linkage to deflect in proportion to the control deflection. This deflection was usually in the opposite direction so as to reduce the control forces. This device would give a constant value of force per g. However, this device suffered from the same disadvantage as all other types of aerodynamic balance, that if the control had to be closely balanced, the forces would be critically dependent on manufacturing tolerances. I devised the method shown in figure 6.2(c) to combine the geared tab with the spring tab so that at low speed the system behaved as a geared tab, but at high speed the spring tab came into play to amplify the pilot's control force, and therefore, avoid the sensitivity of the control forces to manufacturing tolerances. This device was called a geared spring tab. With proper design, a constant value of force per g could be obtained. I published reports on the spring tab and on the geared spring tab to show how they could be designed to give the desired pilot forces. I also determined the limitations of the devices in controlling very large airplanes and showed that for airplanes as heavy as 300,000 pounds (larger than any produced up to that time), the geared spring tab could be used to provide desirable handling qualities. These analyses are combined in an NACA Technical Report (ref. 6.1). A little

thought will allow an appreciation of the remarkable capabilities of spring-tab controls, which allow a human pilot, applying comfortable forces to the control stick or wheel, to maneuver a vehicle weighing as much as 300,000 pounds flying at an airspeed of 300 miles per hour or more.

The spring-tab analyses were discussed with various manufacturers, and spring tabs were widely used on large airplanes. Perhaps the most notable example of their use was on all the controls of the Douglas DC-6, which has had a remarkable record for reliability and excellent handling qualities. Gilruth made the remark that the DC-6 was "the last of the airplanes," meaning it was the last very large airplane in which the pilot, moving the controls directly with his own muscle power, could easily perform all the desired maneuvers. This remark is not completely accurate because other large airplanes of the same period, such as the Boeing B-29, used spring-tab controls. Nevertheless, it illustrates the desire of designers and pilots to have the feeling that they were directly connected to the controls with a reliable mechanical linkage and not dependent on more complex mechanisms such as hydraulic or electric systems.

Spring tabs had many desirable features, but they had two disadvantages. One was a tendency to flutter, which required careful analysis to avoid this dangerous condition. Another was that the tab, deflected in the opposite direction from the control surface, reduced the maximum moment produced by the control. In the case of ailerons, for example, this effect would require ailerons of increased span, which would leave less room on the wing for high-lift flaps. To overcome these disadvantages, a device called a whirleron was proposed by Gilruth. This device consisted of a small windmill mounted on a shaft projecting from the trailing edge of the wing. The windmill was connected to the aileron through gears and shafting so that about two revolutions of the windmill corresponded to full aileron deflection. Normally, the windmill blades were aligned with the wind, but they were connected to the pilot's control stick so that the pilot could control the setting of the blades. The blade angle actually was proportional to the difference between the pilot's input and the aileron deflection so that when the aileron reached the desired deflection, the blade angle would be returned to neutral.

After analyzing this device, a wind-tunnel model was tested. The results appeared favorable when the engineers tested it. However, when the Chief of Research, Floyd Thompson, was invited to see it, he slammed the control stick hard over and the windmill hit the stop so hard that its blades and shafting were knocked loose. Another problem revealed by the tests was that the gyroscopic moments on the blades, which tend to move them into the plane of rotation, caused unstable stick forces. As a result, a second model was built, with blades swept back and counterbalanced to avoid the unstable stick forces, and suitable shock absorbers put in to avoid loads due to hitting the stops. This device worked well, but as will be discussed subsequently, the need for such devices was rapidly decreasing with further airplane developments. Also, no thought had been given to protection against icing.

In running the whirleron tests, I learned something about the motivation of inventors. In connection with the project, I made a library search to see whether any similar work had been done in the past. I found that the British had tested a similar device on a Handley-Page Bomber in the early 1920's. When I showed this report to Gilruth, he commented that they had some really smart engineers back in those days. After that, however, Gilruth never showed the slightest interest in further studies of the whirleron. Perhaps the idea of being the originator of a device is an important factor in determining the inventor's enthusiasm for it.

Research on Closely Balanced Controls

The first really serious compressibility effects on airplanes were encountered in high-speed dives of fighter airplanes in WW II. When pilots put these airplanes in terminal velocity dives starting from about 30,000 feet altitude, they were frequently unable to recover. In cases where the airplane did recover, it often came back with the wings bent or wrinkled due to excessive accelerations in the pull out. The cause of these problems was the formation of shock waves on the upper surface of the relatively thick wings, which caused flow separation. The separated flow, in turn, resulted in loss of downwash at the tail that caused a large diving moment and a large rearward movement of the center of lift on the airplane, which required greatly increased elevator deflection to recover from the dive. The result was that the control force to recover from the dive was greatly increased, sometimes beyond the strength of the pilot. As the air temperature increased with decreasing altitude, the Mach number decreased, which sometimes resulted in rapid recovery of elevator effectiveness, resulting in an excessively violent pull out.

Gilruth proposed a type of elevator control to overcome these problems. The elevator was to be replaced by an all-movable tail surface (now sometimes called a slab tail). This surface was hinged at its aerodynamic center, so that very little hinge moment would result either from deflection or angle of attack. Then, the tail was controlled by a servo tab, which in itself would greatly reduce the pilot's control forces. The result was a tail that required essentially no control force. Since flying qualities studies had shown that pilots required a control force in turns or pull up maneuvers, a weight called a bob-weight (a term coined by the British) was placed in the control system tending to pull the stick forward. This weight would supply a desirable value of force per g in maneuvers. This

force per g would have the advantage of being independent of fore-and-aft location of the center of gravity of the airplane. With conventional elevators, the force per g ordinarily increased with more forward center-of-gravity location, and vice versa, thereby limiting the center-of-gravity range in which satisfactory control forces could be obtained.

An all-movable tail surface of this design was installed on a Curtiss XP-42 airplane (a modification of the P-36) (figure 6.3). After making a high-speed taxi run, the pilot, Mel Gough, came back extremely perturbed and said that the airplane was completely uncontrollable. At first, no one could determine the reason for this unexpected behavior. As a result, I built a tail for a model glider that simulated the design features of the XP-42 tail. When the model was flown, it performed a continuous short-period longitudinal oscillation with the tail banging back and forth between its stops. It was immediately apparent that the aerodynamic center of the tail was actually ahead of the hinge point, which caused the tail to overbalance. Some rather subtle aerodynamic effects were then found to explain this condition, including the floating tendency of the tab, the effect of aspect ratio of the tail, and the effect of the trailing edge angle of the airfoil used. This problem was cured by attaching projecting strips along the trailing edges of the tail. This

device had been used on previous airplanes to increase the hinge moment due to deflection. After this change, the airplane was flyable, but the pilots still considered the control extremely sensitive and unsatisfactory for any kind of practical use.

The cause of the difficulty was analyzed by comparing the stick force variation in a rapid pull up on the XP-42 with that on a similar airplane, the Curtiss P-40, with a conventional elevator. Both airplanes required the same stick force in a steady turn, but in the abrupt short-duration pull up, the stick force on the XP-42 was very small, whereas on the P-40, it was larger than that required for a steady turn at the same value of normal acceleration (figure 6.4). These data are somewhat incomplete because of the short length of the records, but they show quite clearly the large difference in control forces to produce approximately the same value of acceleration in a rapid pull up. The rapid buildup of stick force with the conventional elevator gave the pilot a warning of what value of acceleration to expect if the maneuver were continued, whereas this warning was not present with the XP-42.

Later, to determine whether any characteristics of the all-movable tail itself caused control problems, the tab on the all-movable tail was changed from a servo tab to a geared unbalancing tab. With this arrangement, the

FIGURE 6.4. Comparison of stick forces in rapid pull ups with Curtiss P-40 airplane and Curtiss XP-42 airplane. Both airplanes had the same force per *g* in a steady turn or pull up.

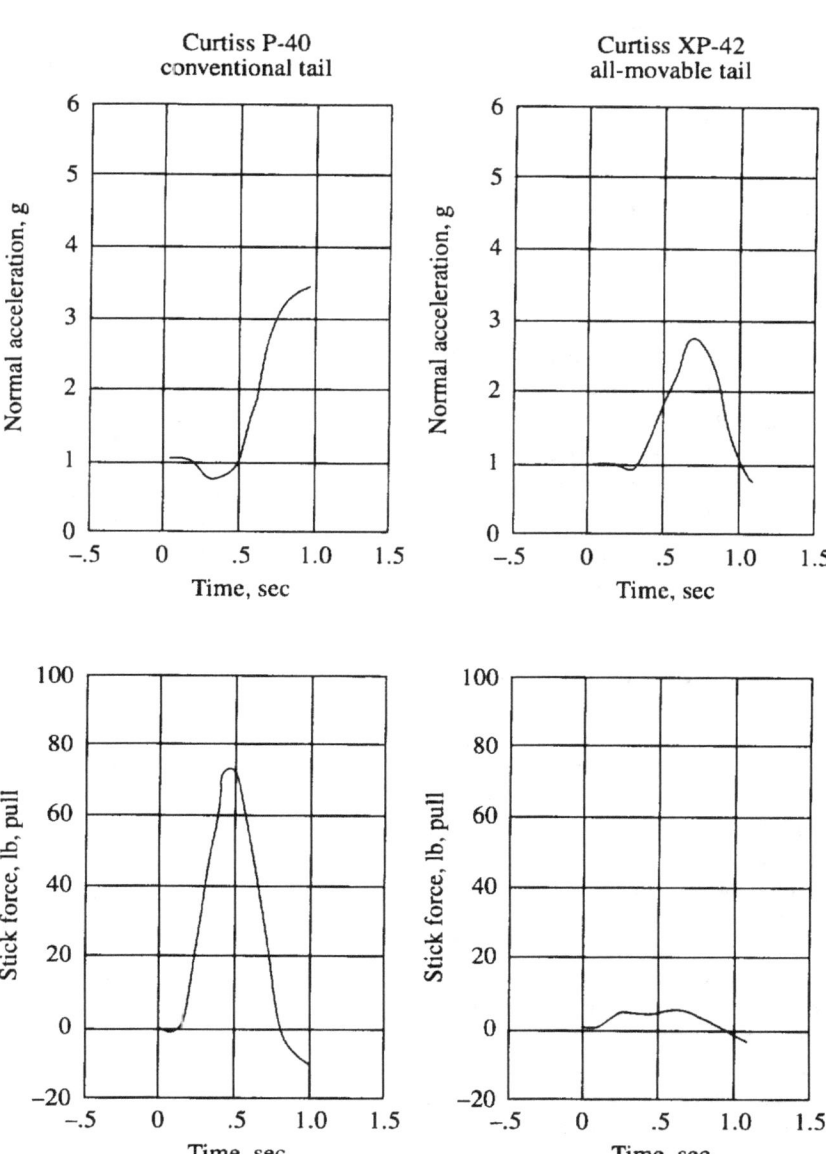

control forces were similar to those on a conventional airplane, and the handling qualities with the all-movable tail were satisfactory.

Evidently what was needed to obtain the desired advantages of the servo tab controlled tail without the problem of inadequate forces in rapid pull ups was a mechanism that would transfer the tab from a

geared unbalancing tab in the first stages of a maneuver to a servo tab as the maneuver continued. This behavior could be obtained by use of a viscous damper in the linkage to allow the tail hinge moment to move the damping piston at a controlled rate, which in turn moved the tab to a position that would result in zero tail hinge moment. This system was tried in the XP-42 with satisfactory results.

After the tests of the XP-42 were started, studies were also made with closely balanced conventional elevators on a Bell P-63 airplane. Again, the control forces were supplied by a bob-weight, and the same problem of lack of warning of the acceleration in a maneuver was encountered. A similar viscous damper installation (also called a dashpot) was installed in series with a spring connected to the pilot's control stick. This system also reduced the undesired variation of control forces with center-of-gravity position and provided satisfactory handling characteristics in maneuvers.

The WW II fighters encountered compressibility effects in high-speed dives, but they were still incapable of supersonic speeds even in dives. With the rapid development of fighter airplane design after the war, however, fighters were developed that reached supersonic speeds, first in dives and later in level flight. At supersonic speeds, the effectiveness of a tab on a tail surface is very small, and the hinge moments of the surface itself undergo erratic variations in passing through the transonic region. As a result, the use of a servo tab to operate the tail as envisioned by Gilruth was not satisfactory. For these airplanes, hydraulic powered controls were developed that provided far more force to actuate the control surfaces than the human pilot could supply. As a result, all the previous emphasis on very closely balanced controls disappeared. The designer was still faced with the problem of providing satisfactory control forces for the pilot, however. In most cases, the force input to the hydraulic actuator was close to zero, and all the pilot's control force was supplied by a feel device.

The design of these feel devices proved to be a difficult problem. In general, most designers first envisioned a bob-weight just as had been tried on the XP-42. Fortunately, the NACA research on this airplane and on the P-63 showed how to design a satisfactory feel device for longitudinal control. I have

always felt that the NACA research on these airplanes was one of the most important contributions to knowledge of handling qualities. As a result, I collected the reports that had been written on these projects in a separate folder and always kept them available for discussions or consultation with designers of new airplanes.

A Control Surface Hinge-Moment Balancing Mechanism

In the previous discussions of control surface hinge moments, the various approaches to reducing the control forces to desirable values have been considered. The earliest technique was use of aerodynamic balance. When airplane speeds and sizes became too large for this approach to work, various types of tab controls such as spring tabs and geared spring tabs were used. When supersonic speeds were reached, tab controls became ineffective and hydraulic power controls were adopted.

In the 1940's during the early stages of this progression, I conceived a fundamentally different idea. I reasoned that if the position of the control surface hinge line could be moved to the point that the lift on the surface multiplied by the distance to the hinge line just balanced the aerodynamic moment on the surface, the surface would be in balance and the pilot's control force would be zero. This idea may be likened to moving the balance point on a see-saw so that it is in balance even though the weights on the two ends are different.

My first analyses considered that a motor or servomechanism would be used to move the hinge. Later, I did some work on the idea that the lift or control force could move the hinge line directly through some aeromechanical mechanism.

The logic for moving the hinge basically depends on the idea that if the lift on the surface is up, then a nose-up moment on the surface should be balanced by moving the hinge line forward, and a nose-down moment should be balanced by moving the hinge line back. If the lift on the surface is down, these motions should be reversed.

A practical problem appears in that as the lift goes through zero, the hinge must be moved from infinity in one direction to infinity in the other direction. In many cases, the lift on a surface does reverse or oscillate in the course of a maneuver. As a result, the mechanism for moving the hinge would have to be very fast.

I first concluded that this solution would be impractical, but as devices for moving controls became more and more complex, I kept coming back to this idea to try different approaches. My last attempt at analysis was made about 1952. I never did reach a solution that appeared at all promising.

One problem that soon appears in attempts at analysis is that the problem is nonlinear if a reversal of lift is taken into account. I did not have the necessary means to analyze such a problem. An analog computer would have made it possible to simulate different types of mechanisms, but such computers were not available at Langley at the time.

Later, when analog computers did become available, it always appeared to me that many of the nonlinear problems that engineers worried about in earlier years were no longer of interest to them and were never studied with analog computers. The excellent analog computer complex at Langley Research Center was discontinued in 1990 because of lack of use. Now, such problems can be studied by digital computers with programs that allow setup of the problem with block diagrams similar to those used for analog computers. The speed of digital computers may not allow real-time solutions of problems involving very high frequencies, such as might have been encountered with the balancing mechanism, though such problems could be studied using an extended time scale.

A Spoiler Operated Servo Control

The servo-tab and spring-tab controls described previously operate well at subsonic speeds, but become ineffective at supersonic speeds because the ratio of control surface hinge moment to lift produced increases greatly under these conditions. The reason for this change is primarily that the pressures on the upper and lower surfaces of the wing caused by control deflection do not carry forward of the hinge line at supersonic speeds. If a tab were used in an effort to defect the control, the tab would have to be so large that the lift produced by the tab would just about offset the lift due to the control surface. To overcome this problem, I conceived a device in which a spoiler ahead of the control surface hinge line produced a moment to deflect the control. This device has also been referred to as a rotary servo tab or a vane-type servo control.

A sketch showing the operation of the device is shown in figure 6.5. A vane or spoiler operated by the pilot is mounted in a tube ahead of the control surface hinge line. When the vane is deflected, its drag causes the tube to rotate. The rotation of the tube is linked to that of the control surface, so that when the vane moves up the control surface also moves up. This device should remain effective at supersonic speeds because the spoiler itself is known to produce a lift force in the correct direction, the reduced pressure behind the spoiler should aid in deflecting the control surface, and the control surface itself would produce lift in the same direction. By use of a feedback linkage from the control surface to the tab input link, the system could readily be arranged to operate in the same way as a servo tab or spring tab.

A model of the device was built and tested by the transonic bump technique (fig. 6.6). The results showed that the control did retain a reasonably large amount of effectiveness at least up to a Mach number of 1, the highest Mach number of the tests (ref. 6.2). A patent

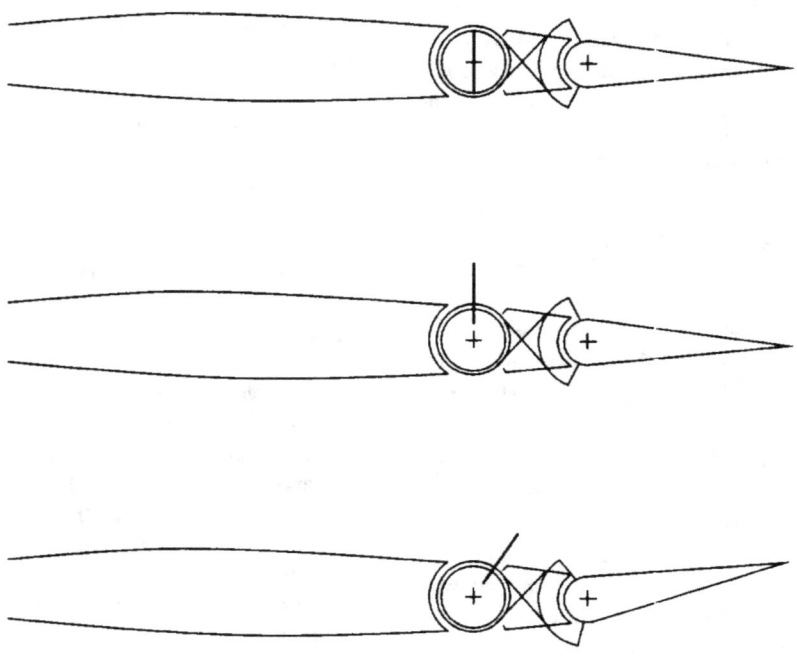

FIGURE 6.5. Diagram of operation of vane-type servo control.

(a) Control in neutral (top).

(b) Vane deflected (middle).

(c) Vane and control surface deflected (bottom).

was obtained on the device, but it was never used on an airplane. These tests were published at a time when the use of hydraulic controls was becoming common, so that the need for such a system no longer existed. This device, however, is probably the only type of manual control ever tested that would retain enough effectiveness at supersonic speeds to control an airplane. The wind-tunnel tests showed that there was considerable buffeting of the control when it was deflected at high speed, which might be a serious problem on a full-scale airplane.

The Introduction of Swept Wings

One of the most important developments in the design of high-speed airplanes during the wartime years was the realization that a swept wing would have much less drag in the transonic and supersonic speed range than an unswept wing. The story of this discovery at Langley is told in *Engineer in Charge*

(ref. 1.1). Though there were initial doubts about the validity of this concept, by the end of the war experimental studies by the free-fall method and other techniques had shown that the concept was indeed correct. In addition, the development of the jet engine had made flight in the transonic range feasible.

Prior to these developments, swept wings were in poor repute for use on subsonic airplanes. It was generally known that the stalling characteristics of swept wings would be adversely affected by the flow of boundary layer to the tips as the stall was approached, which causes a so-called tip stall that results in rapid roll at the stall with no warning in the form of buffeting. In addition, swept wings were poor from the structural standpoint and had lower maximum lift coefficients than unswept wings. As a result, very little experimental data were available on the aerodynamic characteristics of swept wings. The result of this situation was that in practically every wind tunnel, the work in progress was stopped and a program of research on swept wings was initiated.

FIGURE 6.6. Photograph of test model of vane-type servo control. Semaphore-type vanes deflected 12 degrees.

To get some idea of problems that might be encountered with swept wings, I made simplified calculations to determine the magnitude of some effects that might result from this configuration. The dihedral effect, or rolling moment due to sideslip, of a swept wing was known to increase with lift coefficient. In addition to this problem, however, was the effect of the airfoil section pitching moment on the stability derivatives. On an unswept wing, the section pitching moment affects only the longitudinal characteristics, but on a swept wing the pitching moments about axes swept with the wing panels have components that affect the rolling moments. My analysis showed that a swept wing with cambered airfoils would have a reduced dihedral effect. Twist of the tips in the direction to reduce the lift on the tips would also reduce the dihedral effect. These effects were

not then known to most researchers. I presented these results at meetings of the Langley Stability and Control committee.

I also made a simplified analysis of the effect of sweep on the weight and flexibility of a wing. The use of sweep was intended to provide a high critical Mach number, but if the wing had to be made thicker to prevent excessive weight or flexibility, this advantage might be offset. These results were circulated in-house to members of the Structures Research Division, who acknowledged that the results were sound and would have to be considered in selecting the wing span and thickness. Of course, much more detailed study would be required to actually design a swept wing for an airplane. The structural problems that I anticipated have been alleviated on most transports by use of a section

FIGURE 6.7. Bell L-39 airplane incorporating experimental swept wing to study low-speed effects of wing sweep.

near the wing root with a straight trailing edge and by the concentration of stiffness in a spar through the wing root as near as possible to the trailing edge.

Experimental studies of the effects of sweep were also made in the Flight Research Division by methods devised by Robert R. Gilruth. These methods, to be discussed later, were the wing-flow method and the free-fall method. The ready availability of these methods provided much-needed data to the aeronautical industry before the develop-

ment of the slotted-throat wind tunnel made transonic wind-tunnel testing possible.

To get full-scale data at an early date, the Bell Aircraft Corporation also produced for the NACA the L-39 airplane, a P-39 airplane with experimental swept wings (figure 6.7). This airplane was not capable of transonic speeds, but studies were made of the stability and control characteristics and of expected low-speed problems such as poor stalling characteristics. The use of wing slats and other stall-control devices was studied in flight.

CHAPTER 7

Effects of World War II on Research Activities

The research work performed by the NACA at Langley had experienced a slow but steady growth from its start in 1920 to the time I arrived at the center in 1940. The number of personnel had increased from 15 to 739. The research facilities included several wind tunnels and the Flight Research Hangar, which were most all located in the East Area, a section of Langley Field on the side closest to the center of Hampton. I had only a relatively short period of employment before war was declared following the bombing of Pearl Harbor on December 7, 1941. As a result of the war, the aviation industry in the United States experienced a period of phenomenal growth and the research activities at Langley were similarly expanded. The number of professionals at Langley increased from 277 in 1940 to 1158 in 1950, and the total number of employees increased from 739 to 3388. Two new NACA research centers, the Ames Aeronautical Laboratory in Mountain View, California, and the Lewis Laboratory in Cleveland, Ohio, were established just before the war and were managed by personnel transferred from Langley. The war naturally had a major effect on the scope of the work conducted. New technical developments such as the jet engine and the capability of flight at supersonic speeds required many new research facilities and created new technical fields of interest. The administration of the center, however, and the environment for research continued on a surprisingly even keel, probably because the same director and many top management personnel continued in their positions throughout this period. A brief summary of some of the effects of the war on my work in the Flight Research Division is presented in this chapter.

Effect of the War on Living and Working Conditions

When I came to Langley in July 1940, I entered into what seemed like an ideal environment from the standpoints both of living and working. I have described these conditions in more detail in a journal article (ref. 3.1). This was my first experience living away from home and free from the rather rigorous program of studies at MIT. I enjoyed the opportunities for social life, the friendliness of the local people, and the many recreational activities. The model airplane club, with members that included many expert model builders who had been hired by the Langley management from all over the country as technicians and model builders in the shops, gave me an opportunity to meet new people and to continue my hobby of model building and flying.

At the NACA hangar where I worked, some evidence of military buildup was evident. Gilruth had already run flying qualities tests on some of the new military airplanes, such as the Seversky P-35 pursuit plane, the Boeing B-15 and B-17 bombers, and the Vought F4U-1 Corsair Navy fighter. There were still many active projects directed at private and commercial flying. About five light planes, including those built by Piper, Taylorcraft, Stinson, and Bellanca were undergoing tests, and Gilruth had modified a Piper Cub to make it stall- and spin-proof. Ongoing programs were being conducted by the Aircraft Loads Section to make routine measurements of gust loads on airplanes in commercial operations, and the first pressurized commercial transport, the Lockheed C-35, was used to explore turbulence at high altitudes in thunderstorms.

The impact of the war hit suddenly with Pearl Harbor. Hangars were camouflaged and blacked out. For a month or so after the declaration of war, Langley engineers were assigned to guard duty at night in the research facilities to warn of possible terrorist attacks. When I was on this duty, I tried to do a lot of report writing, but I found that the reports written in the loneliness of night tended to be rather rambling and usually required rewriting by the light of day. Soon, regular security guards were hired and engineers were relieved of this responsibility.

The war inhibited social life for two reasons. First, the work week was increased from 44 hours to 48 hours. With work all day Saturdays, there was not much time even for necessary shopping. Also, gas rationing was in effect, which prevented any trips except in the local area. Everyone at Langley worked very hard during the war. This was a period of rapid expansion in which new engineers were hired from engineering schools all over the country. Many of the NACA engineers were sent on recruiting trips, but I did not receive this kind of assignment. Later, starting in 1943, I taught courses to many of the new employees in the subjects of stability and control and handling qualities. These

courses, under sponsorship of the University of Virginia, were given at night at Hampton High School. Preparation of lectures for these courses was a burden on me because I am not a fluent speaker and had to know the course material in detail before I could present satisfactory lectures on it.

One source of concern during the war was induction into the armed services. Some of the Langley employees, of course, felt compelled by patriotism to join the military services. Others had a hard time convincing their local draft boards that aeronautical research work was of sufficient military value to merit deferment. Eventually, the NACA management in Washington, under pressure from the industry to maintain the research and development capabilities at the NACA laboratories, obtained approval from Congress for a plan in which all employees doing essential work at the center would be inducted into the Army then immediately returned to their jobs at the center in civilian status. The details of this arrangement are described more fully in the book *Engineer in Charge*, a history of Langley by Dr. James Hansen (ref. 1.1). I was affected by this plan by having to go up to Richmond for my induction physical, but then immediately returned to my regular work.

Some Effects of the War on Research Activities

Prior to the war, Langley had close relations with the military services through membership on the main committee of the NACA of high-ranking military officials. The projects done by Gilruth on handling qualities, as well as studies by other sections of the Flight Research Division on loads and performance, were strongly appreciated by the military services. As a result, an arrangement was made to supply the NACA with the third airplane off the production line (later changed to the fifth) of every new military airplane produced. This rule was followed for many years. In view of the large number of new military airplanes produced during the war years, this rule supplied Langley and the other NACA centers with an excellent opportunity to contribute to the war effort by recommending improvements as well as by doing research on new problems.

One effect of the war effort was a marked increase in the number of visits by members of the aeronautical industry to discuss the work in progress and a similar increase in the number of trips made to companies and other organizations to assist in the design and development of new airplanes. These visits and trips, which continued after the war, served to solve immediate problems as well as to stimulate the development of aviation through interchange of ideas.

One of the incentives for visits of the NACA personnel to companies was the availability at Langley of recording instruments for flight research. When the Langley Memorial Aeronautical Laboratory first opened at Langley Field in 1920, the director, or Chief Physicist was Frederick H. Norton, a young doctoral graduate straight from MIT. During his brief tenure at Langley, 1920–1923, starting with a staff of seven professionals, he constructed a wind tunnel, started programs in aeronautical and physical research, and established a flight research division. One of his objectives was to design recording instruments to make accurate measurements on airplanes in flight. Dr. Norton went on to become a Professor of Physics and later of Ceramics at MIT. Shortly before Norton's death at the age of 96, Dr. Hansen had a telephone conversation with him, during which Dr. Norton stated that the contribution that he made at Langley in which he took most pride was the development of recording instruments to make scientific measurements on airplanes in flight.

When I started work at Langley, the instruments used were still the same type as those developed by Norton. They incorporated a sensor to detect the physical quantity being measured. The output of this unit was mechanically linked to a small mirror, about one-quarter-inch square, that reflected a beam of light on to a strip of photographic film. The film was moved between a pair of drums by an electric motor. A second mirror was flicked up and down by a timer and reflected another light beam onto the film to give a time reference. A separate instrument was usually required for each quantity being measured. The Instrument Research Division supplied services for calibrating and installing these instruments.

At the time of WW II, the aircraft companies lacked any such facilities for recording quantities in flight. The usual technique was to construct a photopanel, a light-tight box in which a movie camera photographed an array of standard cockpit dial instruments along with a clock to give a time reference. Working up the data from these instruments, frame by frame, was exceedingly tedious.

One of my first trips was to the Republic Aircraft Corporation at Farmingdale, Long Island, New York, accompanying Gilruth and an instrument man. The purpose of the trip was to install instruments and make measurements of the rolling effectiveness of the ailerons on a new airplane derived from the P-47 Thunderbolt (figure 7.1). Later, a Navy pilot flew a captured Japanese Zero to Langley for installation of flight research instruments (figure 7.2). He returned with the airplane to Bolling Field, near Washington, D.C., and made a single flight in which many of the important maneuvers to study flying qualities were performed. I evaluated the film data and published a report on these results (ref. 7.1). Since our instrument technicians usually took several weeks to install instruments in an airplane and the program for measuring flying qualities often required about 20 flights, I was impressed by the speed of the Navy pilot's investigation. For some years after that, I tried to popularize the idea of building a self-contained set of instruments for a quick flying qualities investigation, as well as a regimen of maneuvers that could be used to conduct the investigation in one or two flights. These ideas never caught on, however,

FIGURE 7.1. Republic P-47D-30. This airplane and its later developments were influenced by work at Langley.

FIGURE 7.2. Mitsubishi 00 airplane, popularly known as the Japanese Zero. This captured airplane was instrumented at Langley and tested by a Navy pilot.

because of the availability of the standard techniques and the press of urgent work.

Though Langley was well ahead of the industry in the use of flight recording instruments during the early 1940's, it was apparent to me, after having studied instrumentation under Dr. Draper at MIT, that the instruments themselves were already well behind the state of the art. The instruments had inadequate damping and picked up large amounts of engine vibration in flight. Also, gyroscopic instruments to measure attitude

were unsuitable for flight recording because they usually had no means to disconnect the erection system, which caused errors in accelerated maneuvers. Later developments of magnetic tape recording and the use of multichannel oscillographs to record many quantities on one film were first introduced in telemetering systems used by the Pilotless Aircraft Research Division, but were not available for flight research. It was not until many years later, when such instruments and sensors became commercially available, that the old NACA instruments were finally abandoned.

Returning to the subject of conferences and trips to industry organizations, an interesting point to mention is the large difference in the use of the NACA techniques and assistance by different companies. On the East Coast, Republic and Vought seemed to have close contact with the work in the Flight Research Division, whereas Grumman and Martin were much less involved. An example of the dependence placed by Republic on the NACA capabilities occurred shortly before I came to work. Gilruth was asked to comment on the design of the airplane that later became the P-47 Thunderbolt. Gilruth advised increasing the size of the horizontal tail because of experience with inadequate longitudinal stability on high-powered airplanes with large propellers and the resulting slipstream effects. This advice was taken by Republic, which resulted in a design with noticeably larger horizontal tail than most of the existing fighters. This feature proved valuable later because the airplane had the turbosupercharger and other equipment installed aft of the cockpit, which resulted in more rearward center-of-gravity location than originally planned. Without the larger tail, the airplane would never have had adequate longitudinal stability.

Of course, I am not familiar with the work done in the wind tunnels. Most of the companies at that time did not have their own wind tunnels, so many of them probably employed Langley wind tunnels in tests of their new designs. The Bell Aircraft Corporation, after the war, was strongly involved in the Research Airplane Program. I made several trips to Bell in connection with their proposed designs.

On the West Coast, many companies were more closely associated with the NACA Ames Research Center at Mountain View, California. Still, Douglas representatives occasionally visited Langley to discuss handling-qualities work. Boeing, throughout its existence, appeared to me to be extremely independent. The manufacturer built its own wind tunnel and rarely consulted with Langley on new developments.

Many visits were also made to military installations such as the Naval Test Pilot School at Patuxent River, Maryland, and the Flight Control Laboratory at Wright-Patterson Field. I gave a series of talks to the test pilots at Patuxent on compressibility effects on fighters capable of transonic speeds.

Several cross-country trips were made to conferences at the NACA Lewis Research Center at Cleveland and to the NACA Ames Research Center at Mountain View. In these trips, a group of about 15 engineers would get into the NACA DC-3, which was piloted by NACA test pilots. The trip to the west coast usually required about two stops, one at Memphis, Tennessee, and one in such places as China Lake, California; El Paso, Texas; or Albuquerque, New Mexico. Without these trips, I would never have seen these cities firsthand with the beautiful mountain scenery and the desert climate. I remember in my first trip to the Ames Lab that we stayed in Rickey's Motel in Sunnyvale and had a chance to see the sights of San Jose, such as the Rosicrucian Museum and the Mystery House. I was tremendously impressed by the beautiful city of San Jose, with its gardens and palm trees, and the mountains rising in the distance both in the East and the West. In those days, much of the area was still filled with orange groves. I thought it would be a wonderful place to live. Evidently many other people felt the same way, as San Jose for many years after that was the fastest growing city in the country.

CHAPTER 8

New Ideas—The Fruits of Research

One of the main objectives of a research organization such as the NACA is to come up with new ideas that contribute in some way to progress in aeronautics. It is surprising, however, how few engineers working at such a center ever conceive a new idea in their entire careers. Scientists noted for advancements in their technical fields have stated (and I believe it to be true) that to produce a worthwhile new idea, the researcher must first have a complete knowledge of all advancements that have been made previously. In addition, knowledge of related fields is often helpful because the ability to recognize interrelations between disciplines will reveal new concepts. A second important factor, based on my own experience, is to be involved in experimental work. Such work often requires new techniques, and the old adage "necessity is the mother of invention" certainly applies in this case. In addition, experimental work often produces unexpected results that stimulate thinking on the causes of these results.

Of course, there are some engineers who suggest new ideas that are not worthwhile advancements. They may be ideas that have already been tried or are contrary to accepted physical principles. Such ideas may result from a lack of sufficient background in the field. The possibility of introducing bad ideas is often the downfall of brainstorming sessions in which all members of a group are invited to suggest new ideas in hope of coming up with something useful. Poor ideas are worse than useless, particularly if the proposer is a person with enthusiasm and persistence who is able to convince supervisors that his idea should be developed, often at considerable cost in manhours and facilities.

In my position, I was fortunate in having available the tools for experimental research that required development of new ideas. Some of these ideas were of limited scope and were applicable to the problem at hand or soon became obsolete with the rapid progress of aeronautics. Others had broader application. This chapter presents some examples of new ideas that I developed in my research work.

The First Piloted Simulator Used for Research at Langley

One subject investigated in flying qualities tests was the ability of the pilot to control the short-period yawing oscillations of the airplane. This mode of motion, called the Dutch roll, was usually poorly damped, and with the advent of swept-wing fighter airplanes flying at high altitude, the damping tended to become still less. On actual airplanes, however, it was at that time difficult to vary the

FIGURE 8.1. Yaw chair used for studying ability of human pilots to control lateral oscillations.

damping to investigate the limiting conditions under which the pilot's control would become marginal. To study a wider range of conditions, a simulator called the yaw chair was built (figure 8.1). This device consisted of a chair mounted in a frame free to pivot on a vertical axis. The frame was restrained by springs and was controlled by other springs operated by the rudder pedals. A hydraulic servomechanism was also installed that could apply moments to the chair in phase with its angular velocity, thereby allowing the damping to be varied from stable to unstable. The availability of facilities in the flight research hangar to build devices of angle iron and welded steel tubing was very useful in constructing a simple device of this kind.

The studies made with the yaw chair showed that the pilot could not damp out the oscillations if the period was much less than one second. Effects of the degree of damping or instability were also studied (ref. 8.1). One

interesting result was that the pilot could damp out the oscillations just as well with his eyes closed as with them open. This result showed the importance of so-called kinesthetic cues, which are obtained from the sensory mechanisms in the inner ear, in the control of airplanes by human pilots.

Though the yaw chair was a simple mechanical device capable of studying only one type of control problem, it foreshadowed the development of much more complex general-purpose simulators that became possible with the development of computer and display technology. Operation of such simulators later became a major research activity at the Langley Research Center. Before such developments, however, a simulator called the NAP chair (for normal acceleration and pitch) was built that provided the pilot with a simulation of the short-period vertical motion of an airplane applicable to tasks such as formation flight and in-flight refueling. In the 1950's, a simulator was developed to impart three-axis rotary motion to a cockpit. This simulator had the objective of studying pilots' ability to control lateral oscillations involving large rolling motions. At this time, however, the space age started and the gimbal system ended up in applications to study atmospheric entry or docking. The rotary motion simulator was never used for its intended purpose.

Though these simulators had limited application, they had the advantage of providing very precise reproduction of the motion of the actual vehicle together with visual cues that came directly from the real visual scene. Later simulators frequently used more complex mechanisms to produce the motion of a very heavy cockpit, but because of the mass of the moving system and friction in the drive mechanism they were unable to reproduce the actual motion precisely. Likewise, visual scenes that were produced by television or computer techniques frequently lacked detail of the actual scene or involved lags inconsistent with the motion of the vehicle. Such simulators were often found to be unsuitable for research on the pilot's response characteristics and his interaction with the vehicle dynamics. Simulator engineers were slow to recognize these deficiencies, and when they did, much of the research on the simulators was devoted to improving the simulators rather than on solving real airplane control problems. In recent years, the importance of poorly damped high-frequency airplane motions has decreased because of the widespread use of automatic stability augmentation devices to avoid such characteristics. As a result, many simulator studies deal with problems such as development of instrument displays and training of pilots in specific vehicles. Some modern simulators are used for development of flight control systems. For example, the ability of the human pilot to track a moving target with various types of control systems has been studied. In such applications, consideration must always be given to the sensitivity and speed of response of the human pilot's sensory system as compared to the sensitivity and dynamic characteristics of control force and visual and motion cues of the simulator. A human pilot can track a moving target with an accuracy of better than 0.2 degree. Obviously this problem cannot be studied with a simulator that projects the target with an accuracy of 0.5 degree. In fact, the pointing accuracy of the simulator projection system should be better than 0.02 degree, or one-tenth the pilot's tracking error.

Roll Coupling

Roll coupling means the effect of rapid rolling of an airplane on the motion in other directions, that is, about the pitch and yaw axes. At the time I went to college, such coupling was not a subject discussed in the stability and control courses. In general, airplanes flying up to that time had not experienced problems from this effect in normal flight. It was recognized that in stalls or spins, the interaction of motions about the various axes had to be considered, but analyses of these phenomena had rarely been attempted because of the complex and unknown aerodynamic effects caused by the stalled flow on the airplane.

I would not have tried to analyze the effect of rolling if it had not been for some experimental data that required explanation. During WW II, as mentioned previously, airplanes had started to encounter serious compressibility effects in dives at transonic speeds. Wind tunnels were unable to investigate these problems because of a phenomenon called choking. In a closed-throat tunnel, as the airspeed is raised to the point at which shock waves start to form on the model, a small additional increase in speed will result in the shock waves spreading across the entire test section. Further attempts to increase the Mach number in the test section by applying more power to the tunnel simply result in greater pressure drop across the shock waves with no increase in Mach number.

Gilruth, then head of the Stability and Control Branch, was well aware of the compressibility problems because of the tests being made on high-speed fighter airplanes. He considered how the airplanes being flown in the Flight Research Division might be used to obtain data at transonic speeds. He proposed two new test methods, one called the wing-flow method and the other the free-fall method. The wing-flow method will be discussed subsequently. F. J. Bailey, head of the Performance Branch, was also responsible for many of the ideas required to implement the free-fall method. The free-fall method consists of dropping heavily weighted test models from an altitude of about 30,000 feet. These models would reach Mach numbers of 1.0 to 1.2 before impacting the ground. New developments in telemetering and radar during the war made it possible to obtain useful aerodynamic data from these models during the course of the drop.

The free-fall method was rapidly put into operation, using first a B-29 bomber and later an F-82 fighter, to carry the test models. Actually this was a joint project with the British, in which the British were to develop recoverable models that could be reused to obtain more data, whereas the Americans under Gilruth were to develop expendable models. After about two years of operation, Gilruth's group had successfully tested about 20 models, whereas the British had still to make a successful recovery of a model. The validity of Gilruth's direct approach was thus demonstrated.

Most of the free-fall testing was under the direction of Charles W. Mathews and Jim Rogers Thompson of the Flight Research Division. The models tested were symmetrical bomb-shaped streamline bodies with cruciform tail fins and, in some cases, with cruciform wings. These models fell in a zero-lift trajectory, with measurements confined to drag of various components.

About that time the Bell XS-1 (later called the X-1) rocket airplane was being developed to attempt to reach supersonic speeds.

A question arose as to the effectiveness of the elevator control in the transonic speed range. To study this problem, as well as to investigate the ability of the free-fall method to obtain data in lifting conditions, it was decided to test a model of the XS-1 airplane. A photograph of the model is shown in figure 8.2. The model had a wing span of 7 feet and weighed 1350 pounds. To maintain the desired lifting condition in flight, the model was equipped with a very simple type of autopilot. An accelerometer sensed when the normal acceleration (that is, the acceleration normal to the plane of the wings and fuselage) went outside the range 0.4 to 0.8 g. If the acceleration went outside these limits, the elevator was moved at a slow rate to bring the acceleration back into this range. The model was intended to roll slowly during the drop to maintain a predictable trajectory.

When I fly model airplanes, I customarily look at the model from the rear to check whether the wing is twisted. As a matter of habit, I did this on the XS-1 model. Though the wing was supposed to be machined to an accuracy of one thousandth of an inch, I could readily see that the wing was twisted about one-eighth of an inch at the trailing edge of each tip. Calculations showed that this amount of twist would have caused the model to roll at a rate of about 250 degrees per second during the high-speed portion of the dive. This rate was considered excessive (though we had no way to know, at that time, what rate was really excessive). Therefore two small wedges, like ailerons, were placed on the trailing edge to oppose the built-in twist.

The model was dropped successfully. The exact date of the drop is not known, but a memorandum by C. W. Mathews for H. A. Soulé that gave the results of the drop is dated September 3, 1947. Data from the drop are shown in figure 8.3. The longitudinal control worked as intended. At low Mach numbers, in low-density air at high altitude, the full-up deflection of the elevator of 10 degrees is not enough to maintain 0.4 g.

FIGURE 8.2. Drop model of the XS-1 airplane.

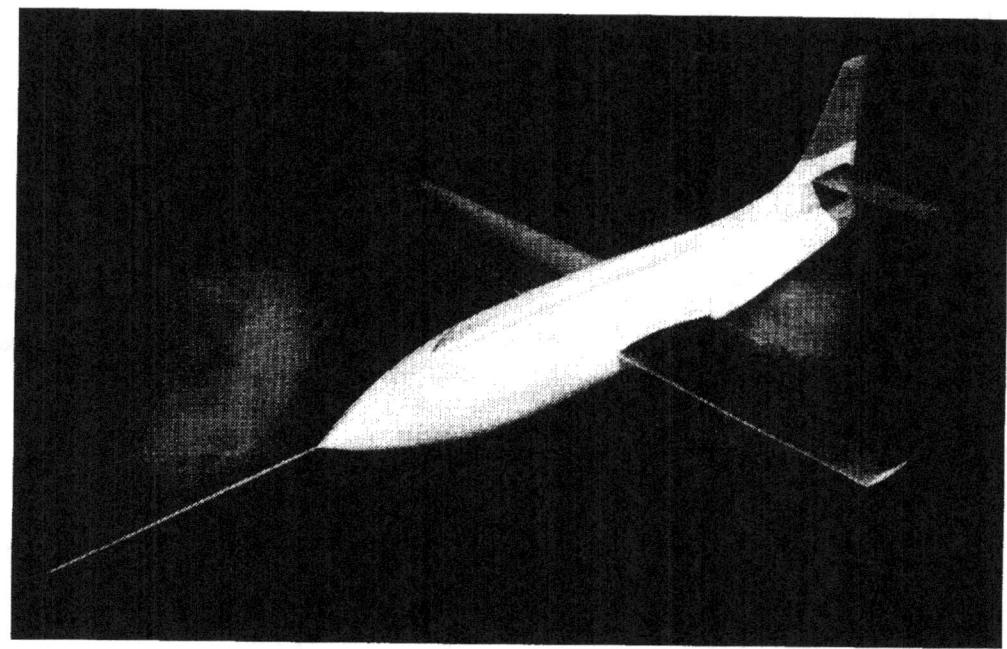

FIGURE 8.3. Flight data from drop test of XS-1 model.

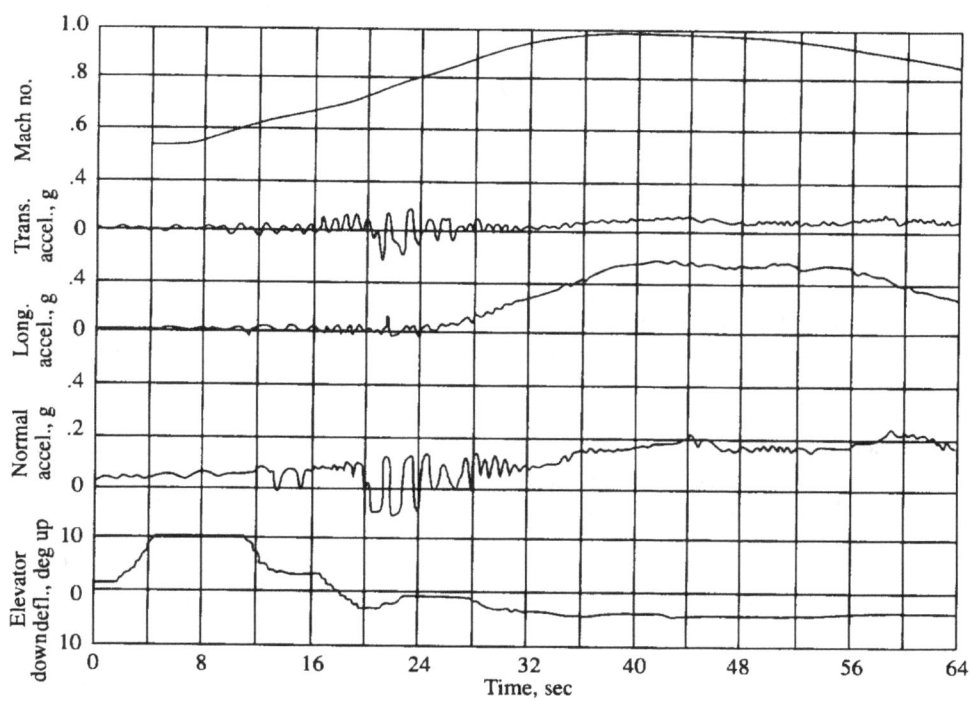

As the speed increases, the model is held in the desired limits of acceleration except for a period of rather wild oscillations at a Mach number around 0.75. At the highest Mach number, full-down deflection of the elevator of 4 degrees is not enough to hold 0.8g, and the acceleration builds up to about 2g.

The model was not equipped with any instrument to measure rolling velocity, but when it was dropped, it was observed in the optical tracker to be rolling by the flashing of light from the wings. Near the end of the flight, however, the model stopped rolling and performed a pull out. It flew far from the designated drop area on a former bombing range called Plum Tree Island and crashed about a mile away in the swamps around the town of Poquoson.

Of course, dropping a heavy bomb in a populated area is not considered very good practice, but evidently no one was aware of the incident. As a result, nothing was ever said about it, and the model, to this day, is probably buried 15-feet deep in the mud somewhere around Poquoson.

In examining the records further, the oscillation mentioned previously was found to represent a violent pitching in angle of attack between the positive and negative stall. The motion below the stall was actually a divergence, but when the stall angle was reached the resulting large increase in restoring moment pushed the model back in the other direction. At this point in the trajectory, the model was well below the Mach number for any serious transonic effects. Knowing the heavy weight of the model and its observed rolling motion, I was led to analyze the problem as some kind of gyroscopic effect. In making the analysis, I used a set of equations called the Euler dynamical equations, which had been taught at MIT by Professor Rauscher as well as in the graduate course Introduction to Theoretical Physics. These equations are nonlinear, but because the methods of solution that I had learned for the airplane stability equations dealt with linear equations, I attempted to linearize the

equations by considering small displacements in pitch and yaw and by assuming the rolling velocity to be constant. I had considerable difficulty in deciding on the signs of the terms involving the coupling effect of the rolling velocity. Finally, I worked out the equations both ways and arrived at the correct sign by considering the simplified limiting case of a rotating rod, the solution for which I could visualize. The resulting equations were also expressed in nondimensional form. Restoring moments were expressed in terms of the natural frequencies in pitch and yaw of the nonrolling airplane, a technique taught for oscillating systems by Professor Draper. In addition, the frequencies were nondimensionalized in terms of the rolling frequency.

When the characteristics of the XS-1 model were substituted in the equations, the results clearly showed the possibility of a divergent motion even at relatively low values of rolling velocity. The instability was likely to occur when the values of longitudinal stability and directional stability were markedly different and when a large amount of the weight was distributed along the fuselage. In the case of the XS-1 model, the directional stability was high, but the longitudinal stability was low, particularly at high subsonic speeds.

This analysis was published in a NACA Technical Note entitled *Effect of Steady Rolling on Longitudinal and Directional Stability* (ref. 8.2). This type of instability had never been encountered on full-scale airplanes, and indeed, it would not occur on the full-scale XS-1. The report predicted, however, that some of the new fighter airplanes then on the drawing board could encounter this problem. In the case of the fighter airplanes, the conditions favorable to instability involved long fuselages, short wings, and high values of longitudinal stability together with low values of directional stability, both of which occur at low supersonic speeds. Later, the report received quite a lot of attention when the problem was actually encountered on airplanes like the X-3 and the F-100. The

FIGURE 8.4. Devices made to illustrate the effect on stability during rolling motion of the equality or inequality of stiffness about the pitch and yaw axes.

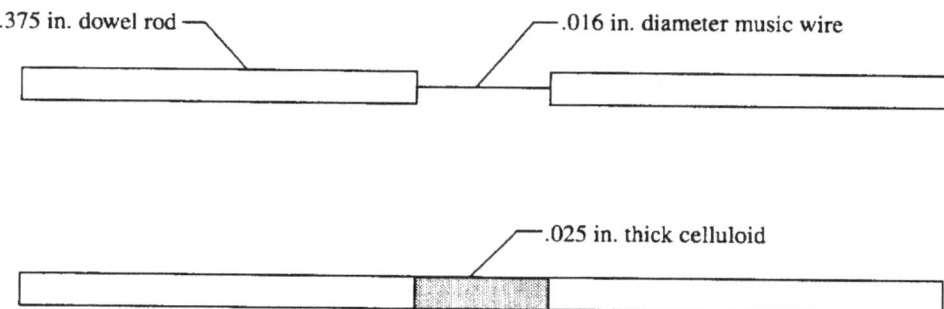

.375 in. dowel rod   .016 in. diameter music wire

.025 in. thick celluloid

existence of my theoretical analysis showed the design changes required to correct the problem and may have saved some pilot's lives by allowing quick action to correct the designs.

Later, the phenomenon was called "roll coupling" or "inertial coupling," and many technical reports were written presenting more detailed analysis of this problem. The more complete analyses took into account the nonlinear nature of the equations and required solutions on analog or digital computers. As a result, the solutions referred to individual cases. I was not sufficiently familiar with these techniques at the time, and the necessary computing equipment was not then available at Langley. My approach was to use the methods that I had learned for the classical airplane stability equations, which were linear. This method revealed the basic physical principles that caused the instability, a result that could not have been obtained by study of more exact solutions for individual cases. I could see that my uncertainty about the sign of the coupling terms could have been avoided if I had started with the complete equations and then linearized them. When others had been able to analyze the complete equations, I felt that my knowledge of this subject was somewhat deficient and should have been treated more thoroughly in the courses at MIT. I realize now, however, that the simpler approach gave a more basic understanding of the problem.

The XS-1 drop test showed that the elevator control was adequate, but in addition

revealed a new type of instability that proved important in the design of future airplanes. This experience shows that the important research results of a test are often quite different from those intended when the tests are planned.

I have received more recognition from the report on roll coupling than from any other work done in my career at Langley. I remember giving a talk on the results at one of the Langley Department Meetings, seminars that were held after working hours to allow different divisions to present their latest research results. I used some crude devices constructed of dowel rods coupled together with either a piece of wire or a flat sheet of celluloid to illustrate the importance of the difference in stability in the two planes. These devices are sketched in figure 8.4. When the rods were spun between the hands, the rod connected with wire would remain stable, whereas the rod connected with the celluloid would fly out to a large deflection. The talk received a round of applause, something quite unusual in these meetings.

This work is also the only research that I conducted that received recognition in the American Institute of Aeronautics and Astronautics magazine, *Astronautics and Aeronautics*, now called *Aerospace America*, which contains a monthly column entitled "Out of the Past," presenting historical items occurring 25 and 50 years previously. Twenty-five years after my report appeared, the magazine in June 1973 presented the following item:

NASA research scientist William H. Phillips of the Langley Memorial Aeronautics Laboratory Flight Research Division publishes Technical Note TN-1627 which contains a theoretical prediction of the problem of inertial coupling. The phenomenon later plagues aircraft having long fuselages and short wings, where the mass of the load is spread along the fuselage with little spanwise distribution of load. In particular, it forces a slow-down in the operational introduction of the North American F-100A Super Sabre, and first manifests itself in flights of the Douglas X-3 research plane. Increasing the area of the vertical tail and the span of the wings solves it.

NACA Research Memorandum H55A13 by the NACA High-Speed Flight Station, 4 Feb. 1955: NASA, Aeronautics & Astronautics 1915–1960, pp. 60–61.

A Method for Determining the Moments and Products of Inertia of Full-Scale Airplanes

To compare flight measurements of airplane motion with theoretical predictions, it is necessary to know the weight, center-of-gravity location, and the moments and products of inertia. In the case of lateral motion, the effect of product of inertia is more frequently expressed in terms of the tilt of the principal longitudinal axis of inertia from the reference axis. At the present time, airplanes, at least those built by large airplane companies, have known weight and inertia characteristics because the companies have a weight control section that keeps up-to-date computer records of the weight and exact location of every item in the airplane. During WW II, however, the weights were less closely controlled, and in many cases the final drawings were not available until long after the airplane was in production. Even today, in airplanes that are fitted with special research equipment or airplanes produced by small manufacturers, weight and inertia characteristics may not be accurately known. For accurate correlation of flight data with theory, therefore, measurements of these characteristics are often required.

Determining weight and center-of-gravity location might appear to be a simple physical measurement if suitable scales are available. Scales for weighing airplanes had been built into the hangar floor, but in years after about 1950, accurate strain gauge load cells were usually more convenient to use. Despite the basic nature of the measurement, new engineers, when assigned this task, often made errors. As a result, a single engineer, William C. Gracey, was assigned the job of overseeing the weighing of each new airplane as it was received. Gracey had also written a report on measurement of the moments of inertia of airplanes by an oscillation technique. This method was occasionally used, but to get moment of inertia in pitch, it was usually more convenient to mount the airplane on jacks and oscillate it while it was restrained by springs of known stiffness attached under the tail.

These methods still failed to supply data on the tilt of the principal longitudinal axis of inertia. A report by Sternfield had shown that the tilt of this axis was of critical importance in determining the damping of the lateral oscillations (ref. 8.3). At the Wright Air Development Center at Wright-Patterson Field, a proposal was once under consideration to build a very large facility that was capable of holding large airplanes and oscillating them to determine their inertia characteristics. So far as I know, this facility was never built. I devised a technique, however, that was at least capable of making accurate measurements of the moment of inertia in yaw and the tilt of the principal axis of airplanes as large as fighters with very inexpensive equipment (ref. 3.4).

FIGURE 8.5. Test airplane as set up for determining moment of inertia and tilt of the principal longitudinal axis of inertia. Spring brackets retouched for clarity.

In this method, the airplane is suspended by a single cable from a crane located in a hangar. Most fighter airplanes are equipped with a hoisting sling to allow them to be moved by cranes or helicopters, so this equipment is already available. The airplane is restrained in yaw by a pair of springs attached to brackets ahead of and behind the suspension cable. From the period of oscillation in yaw, the moment of inertia in yaw can be determined. A photograph of a fighter airplane set up for these measurements is given in figure 8.5.

The pitch angle of a line through the attachment points of the springs can be varied. When the airplane is oscillated in yaw under restraint of the springs, it usually has a rolling oscillation, the amplitude of which can be measured by a position measuring device at the wing tip. For some particular angle of the spring attachment, however, the roll oscillation goes to zero. This point can be found by interpolating between two of the spring settings. The tilt of the principal axis is not equal to this angle at which the roll is zero, but it is related to it by a simple formula

$$\varepsilon = 1/2\tan^{-1}\left(\frac{2I_{Zr}\tan\delta}{I_{Zr} - I_{Xr}}\right)$$

where

ε tilt of principal axis from X body reference axis

δ spring angle for zero roll

I_{Zr}, I_{Xr} moments of inertia about Z and X body reference axes

I have not seen this formula given in any reference book, though I am sure someone must have worked it out previously. So many authors were involved in preparing the brief report (ref. 8.4) describing this project that I did not include my name on it, but I always thought that originating this technique and deriving this simple formula were worthwhile contributions.

Towed Bodies

One piece of apparatus required for most flight tests was the trailing bomb or towed airspeed head. This device was a heavily weighted body containing a static pressure orifice that measured the static pressure sufficiently far from the airplane to avoid errors due to the flow field of the airplane. The pressure was transmitted to the airplane through a tube attached to the towline. This device was used to calibrate the airspeed installation in the airplane, which usually was in error by several miles per hour because of the influence of the flow about the airplane on the pitot-static head mounted on the airplane. These towed airspeed heads had a frequent tendency to become unstable or to break loose and fall to the ground where they could conceivably cause damage or injury. I made a study of the stability of these airspeed heads with the same methods that I had been taught at MIT to study the stability of airplanes. I also studied the previous literature on the subject. A previous study by Herman Glauert, the famous British aerodynamicist, had covered the longitudinal oscillations. I treated the lateral oscillations. My analysis showed that the least stable region was with the body close to the airplane, whereas Glauert showed the instability with the body far from the airplane. I discovered an error in Glauert's analysis in which he had reversed the stable and unstable regions. In view of his reputation, however, I did not mention this discrepancy in my report, which was published in 1944 (ref. 8.5). The Navy was also interested in towed bodies for carrying magnetometers to detect submarines. As a result of their interest, I made tests of a large towed body with a movie camera in its nose to study its motion. Through the course of my career, I always kept an interest in towed bodies and kept a bibliography of reports published on this subject.

Despite my analysis, towed bodies continued to break loose and get lost, and I finally realized that the problem was instability of waves transmitted down the cable rather than instability of the body itself. Studying this type of instability got into the realm of partial differential equations, a subject with which I never had much experience. The problem of transmission of waves down a string or cable, however, is a classical problem in acoustics that was treated in the book, *Theory of Sound* by Lord Rayleigh (ref. 8.6). I studied Lord Rayleigh's analysis that showed how to divide the problem into ordinary differential equations dealing with the time and position dependence of elements of the cable. My analysis showed that oscillations originating from unsteady flow in the region of attachment of the cable to the airplane would be amplified exponentially as they traveled down the cable if the airspeed was greater than the speed of wave propagation. The speed of wave propagation, in turn, varied as the square root of the tension in the cable. With a weighted body on the end of the cable, the tension did not increase very much with airspeed, and as a result, beyond some value of airspeed, the oscillations became unstable. I wrote a second report, which was published in 1949, treating the oscillations of a towed cable (ref. 8.7). After WW II, when much of the German wartime research was made available, I discovered that an almost identical report had been written in Germany (ref. 8.8). These reports revealed the true cause of the loss of the trailing bombs. As a result, another method of static pressure measurement, called the trailing cone method, was developed and has since been widely used. The drag of a cone on the end of the cable increases as the square of the airspeed, and therefore keeps the cable stable as the airspeed is increased. The static holes used to measure the air pressure are placed in the tube attached to the towed cable, sufficiently far ahead of the cone to avoid interference from this source.

I never received much recognition for my knowledge of towed bodies because so few people ever work in this area. Some discussions were held with companies concerned with in-flight refueling because instability of the refueling hose was a matter of concern. I was also selected as an anonymous reviewer of an AIAA paper dealing with the stability of towed cables.

CHAPTER 9

Unsteady Lift

By unsteady lift is meant the lift on a wing or airfoil on which the angle of attack is changing as contrasted to the lift under steady conditions, which exist in steady flight or in most static wind-tunnel tests. The primary application of unsteady lift theory is to wing flutter, inasmuch as on airplanes, even in maneuvers, the angle of attack changes slowly enough that the lift is nearly the same as in steady flight. Other applications, however, are frequently encountered, such as in calculating the response of an airplane to gusts and in analyzing the flight of birds and ornithopters.

When I attended MIT, unsteady lift theory was considered a subject for specialists with a knowledge of advanced mathematics and was not taught in the regular courses. At Langley, a similar situation existed. A few specialists in the Physical Research Division had worked in this field and understood the analyses that had been made, but the average engineer in the wind tunnels or Flight Research Division had no knowledge of the subject.

In my work at the Flight Research Division, I soon encountered problems involving unsteady lift theory. For example, the all-movable tail on the XP-42 airplane could not be flown without some kind of flutter check. Spring tabs were known to be subject to flutter. In gust research, the acceleration on an airplane encountering a gust was used to calculate the gust magnitude producing the response. A correction factor was used based on unsteady lift theory of a two-dimensional airfoil, which I considered to have questionable validity for a finite-span wing. As a result of these problems, from time to time I looked up references on unsteady lift theory and became aware of the studies that had been made in this field.

At Langley, the recognized expert in the field was Dr. Theodore Theodorsen, head of the Physical Research Division, who had written a notable report entitled General Theory of Aerodynamic Instability and the Mechanism of Flutter (ref. 9.1). This report, published in 1935, for the first time put the subject of flutter analysis on a rigorous mathematical basis and gave practical solutions to flutter problems. Incidents of flutter had caused airplane crashes since the earliest days of aviation. For example, Lincoln Beachy, a daredevil stunt flyer who in 1913 flew a strongly braced biplane called the Lil' Looper in air shows around the country, was in a high-speed dive in a show at San Diego when the wings and ailerons fluttered, which caused the airplane to disintegrate and crash. During WW II, Anthony Fokker designed the Fokker D-VIII, one of the first monoplanes with a thick unbraced cantilever wing. The airplane had flown successfully, but a change was made that added weight and stiffness to the rear spar. This change evidently reduced the flutter speed, which resulted in the death of several pilots. Since flutter usually occurs at high values of airspeed and involves the instability of a high-frequency mode of oscillation, the wing extracts tremendous amounts of energy from the air stream that results in a practically explosive increase in amplitude and disintegration of the structure.

Theodorsen's report made a very rigorous analysis of flutter theory. From the standpoint of a beginner, however, this was a poor report to study. Theodorsen outlined the mathematics in very elegant form, with many steps omitted, and gave no references to previous work. In the library, I found that only about four reports had been written on the subject of unsteady lift prior to Theodorsen's report and that someone who completely understood these reports would know about as much on this subject as any expert in the field. (I have in mind the reports by Birnbaum (ref. 9.2), Wagner (ref. 9.3), Glauert (ref. 9.4), and Walker (ref. 9.5)). Unfortunately, I did not have the time or mathematical background to study these reports in detail, but I did find that the analysis by Glauert, the British aerodynamicist, published in 1929 was much easier to understand than the report by Theodorsen. Why, one may ask, was the work by Theodorsen so much more widely acclaimed than that of Glauert? Theodorsen's report did, of course, include the analysis of flutter as well as the theory of unsteady lift. Also, it arrived at the much desired closed-form solution. That is, the results were obtained in terms of Bessel functions. (As shown later by I. E. Garrick, they could also be put in terms of Hankel functions.) Glauert, on the other hand, evaluated the corresponding functions by numerical integration. From the engineer's standpoint, Glauert's results are just as useful, but from the standpoint of the mathematicians who formed the main body of researchers in the field of unsteady lift, the closed-form solution was much more desirable. I did not fully agree with this reasoning, because Bessel functions themselves had been originally evaluated by numerical methods.

The report by Walker, which had been called the greatest doctoral thesis ever written in the field of aeronautics, was fascinating to read and, through use of flow visualization, gave a clearer physical picture than any of the mathematical reports. A point of interest is that Wagner and Walker studied the response of the lift to a step change in angle of attack, whereas Theodorsen and Glauert considered sinusoidal variations in angle of attack. Mathematicians knew that the two approaches should give equivalent results for the linear systems considered, provided the analyses were correct. Later Garrick, who worked for Theodorsen, put out a report verifying that the two sets of results were in agreement. At that time, neither I nor most other engineers knew that the results were equivalent, a fact that further confused the complicated field of unsteady lift.

My own efforts in the field of flutter and unsteady lift were primarily concerned with two projects, a device to reduce the mass-balance weight required to prevent flutter of spring tabs and an attempted clarification of the role of unsteady lift in explaining snaking oscillations. The following sections will discuss these projects.

Spring-Tab Flutter

The spring-tab problem arose because, as shown by Theodorsen, a reliable way to avoid control surface flutter was to mass balance the surface about its hinge line. If this rule was applied to a spring-tab system, both the tab and control surface would have to be mass balanced. Then additional mass-balance weight would have to be used on the control surface to balance the tab mass-balance weight. This weight penalty seems bad enough, but some British reports had shown that the tab mass balance had to be located very close to the tab hinge line to be effective, which means that this balance weight had to be considerably greater than the weight of the tab itself. The reason for this conclusion is shown in figure 9.1. When

FIGURE 9.1. Diagram showing optimal location of mass-balance weight to prevent flutter of spring tab.

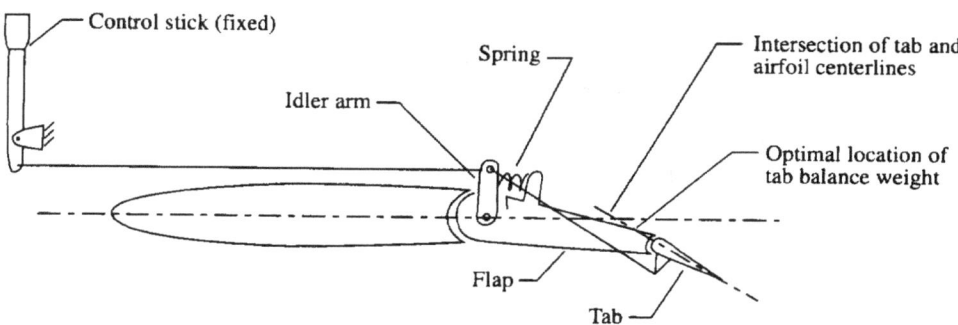

FIGURE 9.2. Schematic diagram of installation of tab balance weight ahead of flap hinge line.

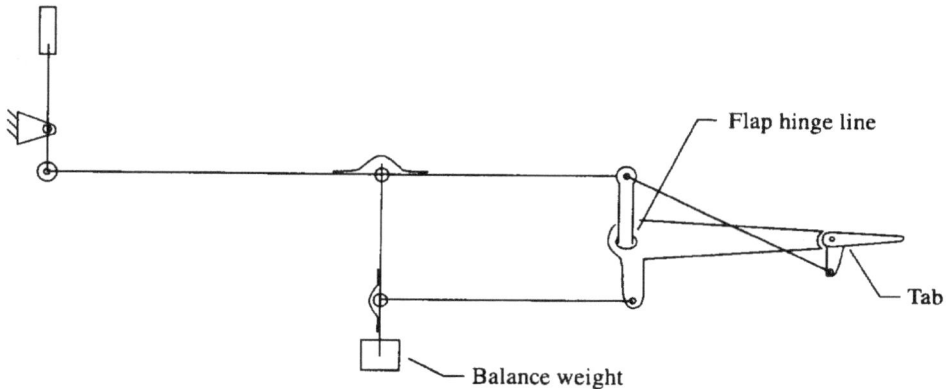

the control surface is moved with the control stick held fixed, the tab ordinarily moves in the same direction as the control surface, but through a greater deflection. This movement results from the kinematics of the tab linkage, which are determined by the desire to have the control follow a quick stick deflection with very little overshoot. For prevention of flutter, an angular acceleration of the control surface should cause the tab to jump ahead of the control surface. A little thought will show that the tab must jump ahead of the position where it would be under static conditions. Referring to figure 9.1, it may be seen that a balance weight located at the intersection of the tab and airfoil centerlines would do no good because it would not move. One located at the tab hinge line would be ineffective because it would exert no moment about the tab hinge. The best location for the mass-balance weight is half

way between these two points. The result of the heavy tab balance weight and the added weight of the main control surface balance to balance the tab balance weight was an undesirably large weight penalty. I devised a linkage arrangement in which the tab balance weight could be located ahead of the main surface hinge line, where it did not add to the weight behind the hinge line and yet would cause the tab to jump ahead of its static position in a sudden control movement. A drawing of the linkage is shown in figure 9.2. A memorandum was written for the Engineer in Charge, Dr. H. J. E. Reid, proposing this device. As a result, a discussion was held with Dr. Theodorsen, who was familiar with the problems of tab flutter, but who did not appear to know of the recent British reports on the subject. He felt that the device would prevent low-frequency flutter, but might be ineffective in preventing flutter of some

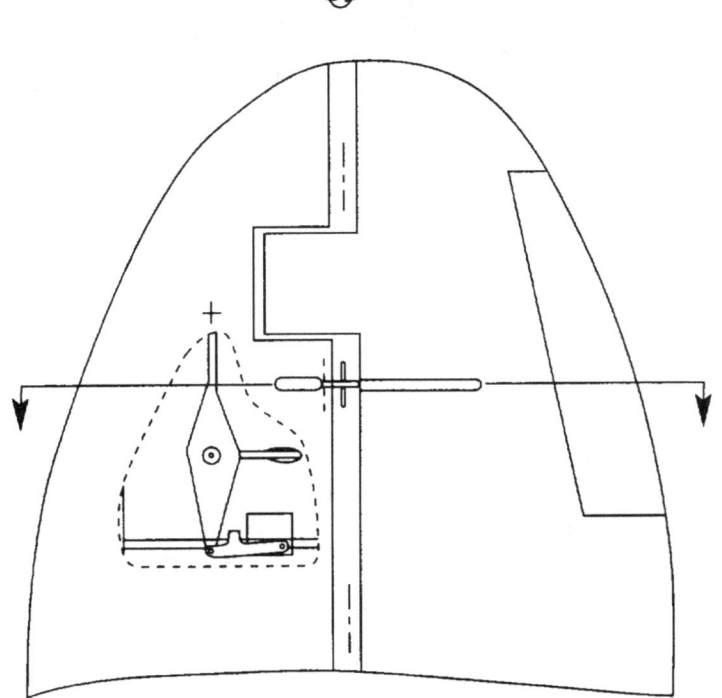

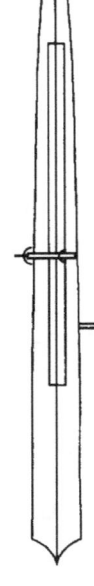

FIGURE 9.3. Drawing of balance weight installation in B-34 vertical tail.

high-frequency modes involving flexibility of the structure or the linkage. He proposed building an installation of the system in an actual control surface to allow measurement of the high-frequency modes.

With the endorsement of Dr. Theodorsen, therefore, a vertical tail was obtained from an Air Force B-34 bomber (similar to the Lockheed Hudson) and the system was installed (figs. 9.3 and 9.4). This tail surface was suitable for installation in a wind tunnel for testing as part of an ongoing series of flutter tests.

When Theodorsen was invited to the hanger to see the installation, he hauled back with his fist and hit the tail a solid blow, which caused it to shudder and shake. His comment was, "It will flooter!" Later vibration tests were made, which showed that there were resonant modes at 14 and 25 hertz in which

the balance weight moved oppositely from its motion at low frequencies. Most of the flexibility was in the thin-skinned structure of the tail, whereas the linkages themselves were quite rigid. As a result of these tests, it was concluded that the device might be subject to high-frequency flutter, and no further studies were made.

I have heard recently that the elevator on the Curtiss C-46 airplane did have a device similar to the one I tested to move the tab balance weight ahead of the control surface hinge line, though most actual spring-tab installations on service airplanes did not have balance weights on the tab, and the main provision to avoid flutter was to make the tab as light as possible. These installations were tested in flight at gradually increasing speeds to verify that they did not flutter within the desired flight envelope. It therefore appears

FIGURE 9.4. Photograph of balance weight installation in B-34 vertical tail.

that the British analyses requiring the use the balance weights for absolute prevention of flutter were unduly conservative. Also, as mentioned previously, the need for spring-tab controls rapidly disappeared with the introduction of hydraulic powered controls.

Effect of Unsteady Lift on Snaking Oscillations

A peculiarity of the two-dimensional unsteady lift theory of Theodorsen and other investigators is that the curve showing phase angle of the lift as a function of frequency in an oscillation of angle of attack initially has an infinite negative slope. This result means that at very low frequencies of oscillation,

the force in phase with the velocity of the motion is much larger than the force in phase with the displacement. If the surface were used as a vertical tail, for example, this theory would indicate that in an oscillation of low frequency, the tail itself might tend to make the oscillation unstable.

The prevalence of snaking oscillations on many airplanes, as mentioned previously, made it tempting to look for some basic explanation of the oscillation based on aerodynamic theory. Reports by Smilg (ref. 9.6), a flutter expert at Wright-Patterson Field, and by Pinsker (ref. 9.7), a well-known aerodynamicist at the British RAE, made it clear that serious thought was being given to this explanation. Personally, I was well satisfied with the explanation involving control system friction that has been discussed previously, and I felt that it was incorrect to apply two-dimensional theory (that is, a theory that applies to a wing of infinite span) to the case of a low aspect ratio vertical tail. Proving this belief, however, was a difficult problem.

A remarkable report had been published in 1940 by Robert T. Jones on *The Unsteady Lift of a Wing of Finite Aspect Ratio* (ref. 9.8). Jones had avoided the complex mathematics of previous investigators by considering the response to a step change in angle of attack, then converting this result to the response to sinusoidal oscillations. To make this conversion in a simple way, Jones fit the variation of lift following a step input with an exponential curve. I was not sure what effect this approximation would have at low frequencies; therefore, I attempted to make an analysis similar to that of Jones, but working from the start with a sinusoidal input.

Today I would have less concern about the accuracy of Jones' results, but at that time I was less familiar with the transformation between step and sinusoidal inputs. I hoped to be able to make an analysis in which the mathematics, at least, was exact, thereby obtaining a direct basis of comparison with Theodorsen's two-dimensional results.

The basis of the analysis in Jones' theory was to convert the vortex system of an infinite wing, which consisted of a bound vortex and a starting vortex, to that of a finite wing by superimposing a pair of horseshoe vortices that may be called "canceling vortices." These vortices cancel the parts of the vortex system beyond the wing tips and add the trailing vortices of the finite wing. I used the same technique, but allowed all these vortices to vary sinusoidally with time and position. The effects of each element of each vortex on the downwash on the wing was obtained by integration. The resulting integrals were evaluated with the help of Keith Harder, a mathematician in the Physical Research Division. All the integrals could be found in tables of integrals except one term of one integral. This term was evaluated by

graphical integration. As a result, I did not realize my hope of obtaining a closed-form mathematical solution.

In 1952, the results were put in the form of a report in which my analysis and the work of various investigators in the field of unsteady lift of finite wings were compared, and the effect of these results on lateral oscillations of airplanes was demonstrated. The report was reviewed by an editorial committee consisting mainly of members of the Physical Research Division, including mathematicians who had devoted much of their careers to studying unsteady lift effects. These mathematically inclined individuals were much more concerned with the mathematical development than with the application to airplane oscillations. They concluded that the theory really did not contribute any new results in unsteady lift theory, and therefore they recommended that the report should not be published. I was disappointed that the paper was not published, mainly because I felt that the two-dimensional unsteady lift theory had been frequently applied incorrectly in the past. Also, I felt that my report would make the subject more understandable to the average engineer.

An important point in the presentation of my results was the method of expressing reduced frequency. In previous reports on unsteady lift theory, the frequency had been expressed in nondimensional form by use of the parameter $\omega c/V$, where ω is the frequency of the oscillation in radians per second, c is the wing chord, and V is the airspeed. I found that at low frequencies, the parameter $\omega b/V$, where b is the wing span, was much more suitable. This parameter would be meaningless in the two-dimensional case because the span b would be infinite. In the case of the finite wing, however the results showing the phase lag of the lift was nearly independent of aspect ratio where the reduced frequency based on span was used. This result shows that at low frequencies, the trailing vortices, which disappear in the two-dimensional case, remain important and, at low frequencies, are mainly responsible for the phase

FIGURE 9.5. Amplitude and phase angle of the circulatory lift of elliptical wings resulting from a sinusoidal change in angle of attack, represented as the vector C_{L_α}, as functions of $\omega b/V$, where ω is the frequency in radians per second, b is the wingspan, and V is the airspeed.

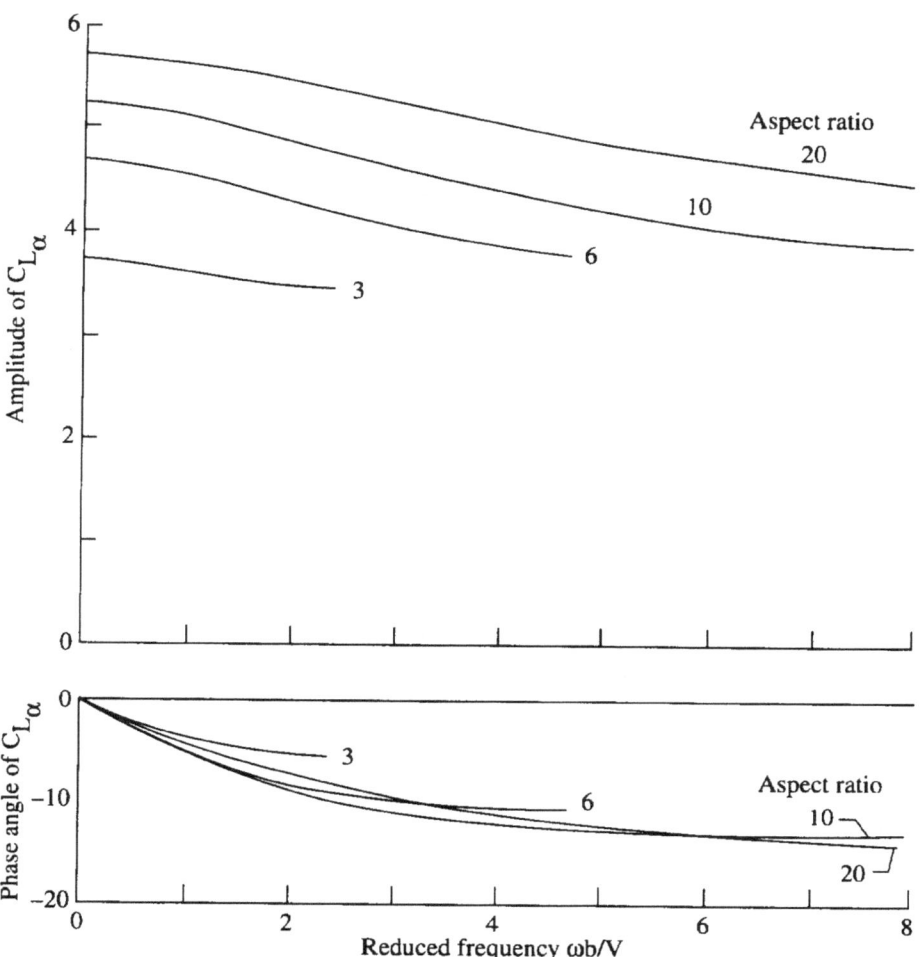

variations of the lift. The results indicate that the slope of the phase angle curve remains finite as the frequency approaches zero, though my analysis, because of the use of graphical integration, did not demonstrate this point rigorously.

To illustrate these points, figure 9.5 shows the variation of the amplitude and phase angle of the circulatory lift with frequency based on wing span. Note that the two-dimensional case can not even be shown when the frequency is based on wing span. The damping of oscillations based on two-dimensional and three-dimensional theories are shown in figure 9.6. The damping based

on the three-dimensional theory is almost the same as that based on conventional stability theory, called the quasi-static case, which considers the angle-of-attack change at the tail due to yawing velocity, but then neglects unsteady lift considerations altogether. The damping based on two-dimensional theory, on the other hand, is considerably less over the whole range of frequencies shown.

One beneficial result of this project was that I obtained a working knowledge of unsteady lift theory. I prepared a talk on this subject that minimized the mathematical aspects. This talk was presented on several occasions to engineers working in my division. Since

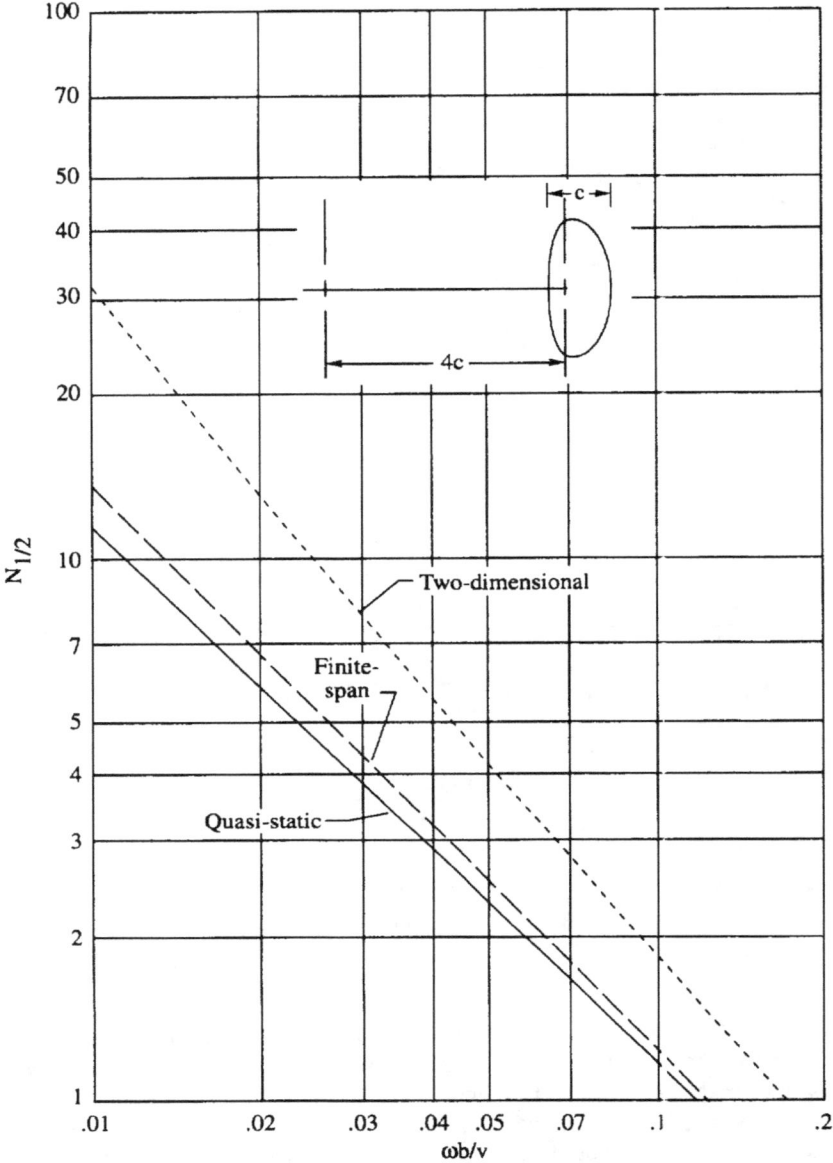

FIGURE 9.6. Damping of oscillations of a weathervane computed by finite-span, quasi-static, and two-dimensional theories. $N_{1/2}$, the number of cycles to damp to half amplitude, is plotted as a function of $\frac{\omega b}{v}$, where ω is frequency in radians per second, b is span of weather vane, and v is velocity. Frequency varied by changing the inertia.

this work was completed, the use of high-speed computers has greatly extended the work on unsteady lift, which allows more accurate calculations for wings with various planforms and mode shapes at speeds throughout the Mach number range. This field still remains a subject for specialists, though some joint programs have been conducted in recent years to combine the efforts of control theorists, structural dynamicists, and unsteady lift specialists to allow applications of control theory to predict flutter of a flexible structure and to design automatic flutter damping systems.

CHAPTER 10

An Automatic Aileron Trim Device for Personal Airplanes

This subject is accorded the recognition of a separate chapter because of the depth of my personal interest in the subject and because of the long period of time over which the problem of spiral instability had repeatedly come up for consideration. In a previous section of this paper, under the heading "Spiral Instability," the tendency of airplanes to diverge from a straight path into a spiral dive was noted. This divergence was said to be so slow that it is rarely noticed by the pilot when he has visual reference to the horizon. If the airplane is flying under instrument conditions, a pilot who is proficient in instrument flight likewise has no difficulty with this type of instability. Many crashes have occurred, however, when novice pilots flew into clouds or fog banks and were unable to keep the airplane flying level. This problem becomes more severe in light airplanes because they are usually equipped just with the basic instrumentation (turn and bank indicator and airspeed) which require considerably more skill to fly blind than an airplane equipped with an artificial horizon.

I was profoundly concerned with this problem because of the experience of one of my professors at MIT. The phenomenon of spiral instability and its associated dangers had been discussed by Professor Otto Koppen in his courses on airplane stability and control. He was a pilot himself, and he took me and all his other students up in a Fairchild 24 to give demonstrations of all the stability characteristics discussed in his courses. Shortly after I left MIT, Professor Koppen's sixteen-year-old daughter and her instructor, who was giving her flying lessons, were both killed when their small airplane crashed after flying into a fog bank. Professor Koppen, of course, was devastated and was unable to teach for two years after that.

Development and Tests of the Trim Device

I knew that the problem of spiral instability could be cured with a conventional autopilot, but the existing autopilots were too expensive and heavy to be used in light airplanes. Much of the complication and expense of conventional autopilots resulted from the need to operate the controls in proportion to various signals sensed by instruments in the airplane. This process, in those days, required the use of vacuum-tube amplifiers and complex servomechanisms for converting small electrical signals to mechanical outputs. I therefore gave much thought to a simpler way to perform this function.

I had been interested in bang-bang controls (also called discontinuous controls) in which the signal to operate the control was switched either full on or full off. With this type of mechanism, an on-off contact replaces the complex electronic amplifier. Such systems can be analyzed by the phase plane method that was described in the book by Minorsky on nonlinear systems (ref. 5.5). Preliminary studies of wing-leveler controls were made by using this technique.

At that time, about 1955, the electronic analog computer had recently become available. This computer greatly aided the study of nonlinear systems. One of the engineers in my branch, Helmut A. Kuenel, set up the problem on an analog computer to study various ideas in more detail than could be done using the phase plane method.

As mentioned previously, most airplanes, whether or not they are inherently spirally stable, will diverge from straight flight because the friction in the control system is large enough to hold the controls deflected from the trim position. To try to overcome this problem, a previous flight study had been made in which preloaded centering springs could be connected to the aileron control linkage to overcome the friction in the system. The spring device could be adjusted to the proper trim position by the pilot. This device prevented the airplane from diverging when it was properly trimmed, but it could not take care of subsequent changes in trim caused by asymmetric fuel usage or by changes in power or airspeed. Evidently, a device was needed to automatically move the centering spring unit to the correct trim position.

A gyroscope was needed to sense yawing velocity (turning) of the airplane and move the centering spring unit as required to cancel this motion. A bang-bang control, in which a slow-moving electric motor connected to the spring unit would be switched to run in one direction or the other, appeared ideal for this purpose. In analytical and analog computer studies, however, the device was shown to need some anticipation of the return of the springs to the trim position, otherwise the device would overshoot the trim position and cause a slow oscillation with the

airplane turning first one way and then the other.

Several methods of providing this anticipation (called a lead signal) were tried, all of which were more complicated than desired. Finally, the idea was conceived of simply tilting the spin axis of the gyroscope so that it would sense components of both rolling velocity and yawing velocity. Since the rolling velocity as the airplane rolls into a turn leads the resulting yawing velocity, this method provided the desired lead signal. This method was so much simpler than any of the other techniques that it was immediately adopted.

A suitable gyroscopic device was provided by H. Douglas Garner of the Instrument Research Division. This device, taken from other equipment, already incorporated a pair of contacts on the gyroscope gimbal to provide the desired operation of the trim motor. In addition, it had a means of electrically torquing the gimbal to bias the rate of turn at which the contacts were centered. Mr. Garner later provided valuable ideas on use of this feature in the system. The gyroscope was mounted on a tilting platform and the system was installed in a Cessna 190 airplane available at the NACA Flight Research Division.

The development of the system and results of flight tests are reported in detail in a NACA report (ref. 10.1). A diagram of the system taken from this report is given in figure 10.1. The system provided excellent stabilization of the airplane in smooth or turbulent air. It worked even better than expected as a result of two fortuitous features. First, with the gyroscope set to the correct tilt angle, the airplane, when released from a banked attitude, returned to level flight in a practically deadbeat manner, with little or no overshoot. It would be expected that a different gyroscope tilt angle would be required for each value of airspeed, but it was found (and shown analytically) that the variation of angle of attack of the airplane as it flew at different values of airspeed was in the correct direction, and

FIGURE 10.1. Schematic diagram of automatic aileron trim control device.

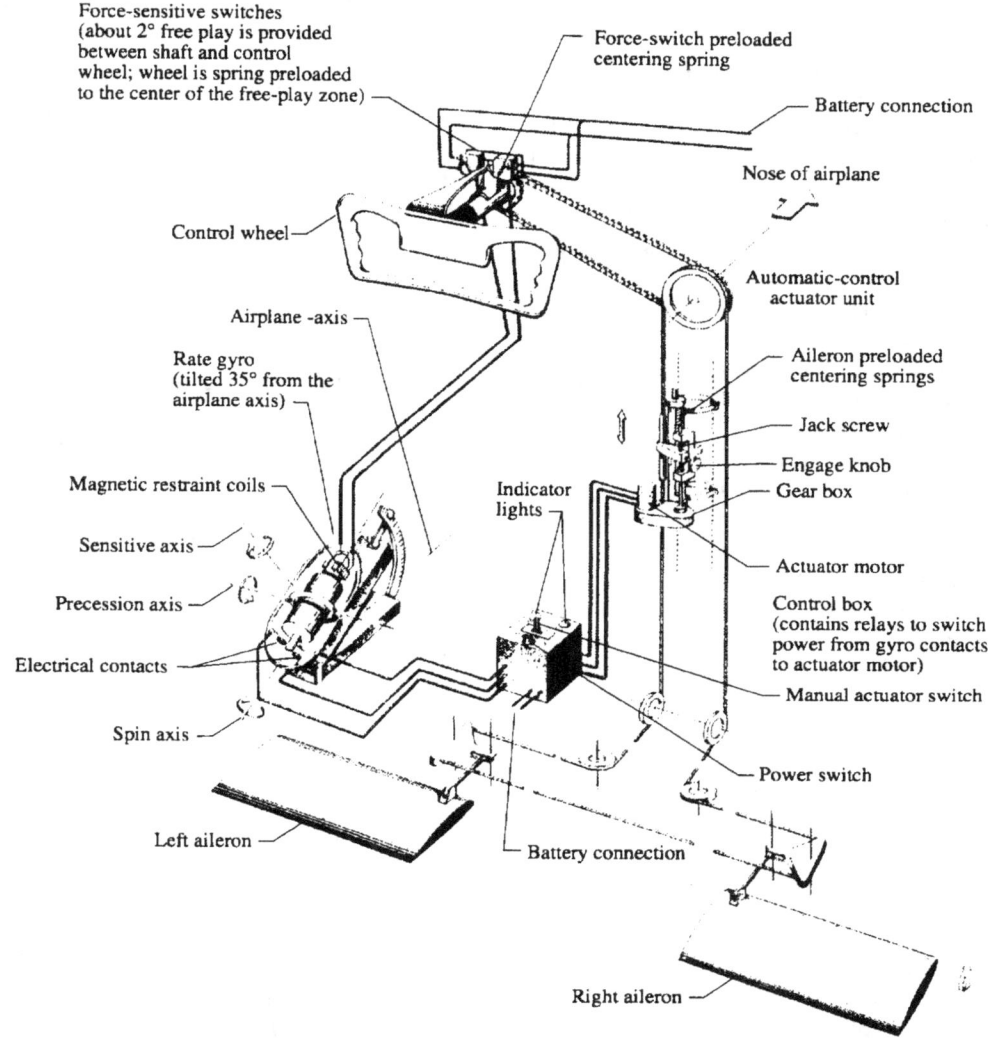

approximately of the right magnitude, to provide automatically the required variation with airspeed of the tilt angle from the flight path. Second, any bang-bang, or discontinuous, control would be expected to have a small hunting oscillation about the trim position. In the actual installation, however, the engine vibration caused the gyroscope contacts to chatter as the airplane reached the level attitude. As a result, the action of the control approximated a linear control and the hunting oscillation was eliminated. This system then behaved as a "dual mode control" with a discontinuous action at large displacements and a linear action at small displacements. This type of system had been advocated by some control theorists to take advantage of the best features of both types of control, but it had probably never been implemented in such a simple manner.

A problem observed by the pilots was that if the airplane was placed in a steady turn and then manually returned to level flight on a desired heading, the control system, operating more slowly than the human pilot, would leave the ailerons deflected and would then continue to stabilize the airplane on a heading a few degrees displaced from that desired by the pilot. To avoid this problem, H. Douglas Garner suggested that a pair of force switches be placed on the control wheel between the pilot's grip and the wheel itself, which would apply current to the torquing coils on the gyroscope gimbal. With this system, as the human pilot held the airplane on the desired heading, the airplane would be retrimmed on this heading by the automatic system.

The force switches also allowed the pilot to make constant-rate turns of about three degrees per second for instrument flight maneuvers by holding a small force on the control wheel within the preload of the centering springs. The system therefore provided features of a more sophisticated autopilot in addition to the original objective of acting simply as a wing leveler.

Because of the preload in the centering springs, the human pilot had to apply a control force exceeding this amount (about four pounds on the control wheel) to maneuver or to hold a steady turn. The NACA test pilots objected to this feature because of their experience with airplanes that did not oppose the pilot in holding a banked attitude. Beginners with little piloting experience, on the other hand, generally preferred the feeling that the airplane would return to level flight if the controls were released.

A patent disclosure for the system was filed. At that time, all the NACA patent applications were handled by one man at the NACA Headquarters in Washington. About a year later, the formal NACA report (ref. 10.1) was published. I did not know at the time that a formal report placed the invention in the public domain and that the patent application was thereby voided. I was displeased that a year had passed without action by the Headquarters office to file a patent application, but

I did not worry about it because I had no interest in making money from the invention.

Later, when the report became known among light airplane companies and control system manufacturers, about eight of these companies sent representatives to Langley for further discussion of the device. The first question asked usually was, "Is the device covered by a valid patent?" Without a patent to give an exclusive right to produce the device, none of the manufacturers would consider marketing it. As a result, I learned the value of a patent even in cases when the inventor was not interested in making money. Several light plane autopilots were in production at that time and continue to be available today, but because they are expensive and require considerable rebuilding of the airplane for installation, they are rarely used. These problems would have been avoided by the system tested. Mr. Garner has continued research in this field and has produced designs for autopilots suitable for construction and installation by home builders. These devices further extend the capabilities of the device tested by providing a heading hold mode. These devices are highly successful, but unfortunately many light airplanes still are not equipped with autopilots and crashes in instrument flight conditions continue to occur.

The wing-leveler autopilot described is notable because, to my knowledge, it is one of the few successful applications of a discontinuous control system for airplane control. Discontinuous controls have been widely investigated by control theorists. This type of control is theoretically more efficient in correcting for disturbances because it uses the full power of the servomechanism at all times. A control system that uses its maximum effort in correcting for a disturbance that is followed by maximum effort in the other direction to stop the motion at the desired point has been called an optimal bang-bang control, because it can bring the system to the desired point in a shorter time than any other system (for example, a linear system) with the same maximum control effort. Despite this theoretical advantage, these systems have found little use in aeronautical applications, though they have sometimes been used for spacecraft attitude control. In most cases, the tendency to hunt back and forth in a limit cycle oscillation after correcting for the disturbance has been a serious drawback. The availability of computers makes the study of such systems much easier. In recent years, a group of researchers from a British university came to Langley to discuss their work in this field. They were very surprised to hear that a successful application of a discontinuous control had been made forty years ago.

To conclude this account, the subsequent career of Professor Otto Koppen might be of interest. After the death of his daughter, Koppen eventually went back to teaching at MIT, but his wife made him promise that he would never fly an airplane after that. As long as she was alive, he kept this promise, but after her death he took up flying again. A few years ago, I learned from someone in the Langley Flying Club that Professor Koppen had been observed flying over Langley Field on his way to Florida in an American Yankee, a very small high-performance personal plane. At the time, he must have been in his 80's. Later, I learned that Professor Koppen was the oldest pilot in the country with a full instrument rating and that his plane was equipped with a wing-leveler autopilot and special navigational instrumentation, to the extent that there was no room in the cockpit for anyone but the pilot. He died in 1993 at the age of 96.

CHAPTER 11

New Areas of Research

Many of the disciplines that I studied in college were useful in my work at Langley, as has been shown in the preceding chapters. Some other subjects of interest, however, were completely new and resulted from the rapid progress of aeronautics during the war years. These topics included compressibility effects, that is, the effect of approach to the speed of sound on the flow around aircraft. The word "compressibility" is derived from the concept that at speeds attained by aircraft prior to WW II, the air flowing over the airplane may usually be considered incompressible because the local changes in pressure or density of the air caused by the passage of the airplane are very small compared to the value of these quantities in the ambient atmosphere. As the airplane flies at a speed approaching the speed of sound, however, these quantities experience changes that are no longer small compared to the ambient conditions. As a result, large changes in flow patterns and aerodynamic characteristics occur.

Another development requiring new approaches was the necessity of hydraulically powered controls in the primary control systems of airplanes. Such systems, as shown previously, were required because the conventional means of operating the control surfaces were ineffective in the transonic and supersonic speed ranges.

Finally, the developments in control theory that occurred during the war allowed studies not only of automatic pilots, but also of the control actions of human pilots in ways that had not been considered previously. Because of the strong dependence of airplane behavior on the actions of the pilot, this subject attracted a lot of attention and became a whole new field of research.

The Wing-Flow Technique

The problem of compressibility effects during dive recoveries of high-speed fighter airplanes during WW II has already been mentioned. In chapter 8 in the section entitled "Roll Coupling," my branch head Robert R. Gilruth was said to have developed two techniques to study compressibility problems, the free-fall method and the wing-flow method. This section describes the wing-flow method and some of the research conducted by this technique. I was not responsible for the invention of this concept. It was typical of Gilruth's approach to research, however, that he did not hesitate to assign capable engineers to work on ideas that he considered important, even though these studies might be quite different from the traditional work of the group. As a result, my section, the Stability and Control Section, was

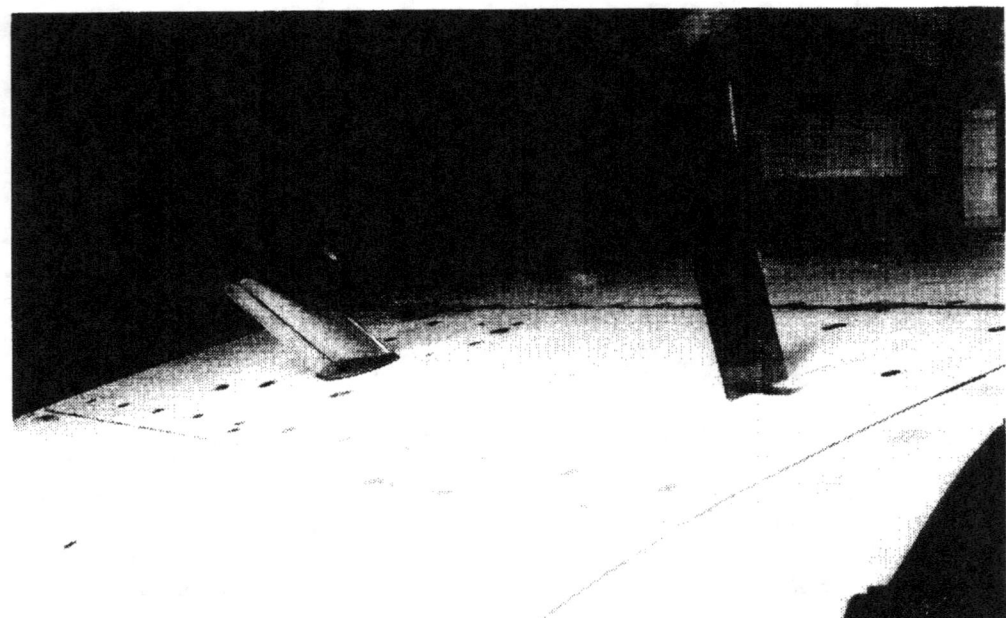

FIGURE 11.1. Wing-flow model and test apparatus mounted on P-51 wing. A flow-direction vane is also mounted outboard of the test model.

assigned the job of developing the wing-flow technique along with members of the Performance Section, and I was in charge of supervising the design of equipment and conduct of some of the tests.

In the wing-flow method, a small model is mounted on the upper surface of the wing of a fighter airplane. When the airplane dives from a high altitude and a pull out is made at a high but still safe Mach number, the flow on the upper surface of the wing smoothly increases from subsonic to supersonic speeds. A region of reasonably uniform flow exists that may be used as the test region for a small model. The choking phenomenon occurring in closed-throat tunnels does not exist because of the lack of boundaries above and at the sides of the test region.

Figure 11.1 shows a photo of a small model mounted on the wing of a P-51 airplane. A special glove was built on the wing to give a more uniform flow region, and the velocity field in this region was measured in dives before installing the model. As the airplane goes through its dive and pull out, the model is oscillated back and forth at a frequency of

about one cycle per second to vary either angle of attack or flap deflection. The forces on the model are continuously recorded with a strain gauge balance and a recording oscillograph. The dive lasts about 30 seconds and in this period the Mach number at the model increases from about 0.7 to 1.2. A drawing of an unswept semispan wing-flow model of a rectangular wing with a full-span flap is shown in figure 11.2, and some of the data obtained are shown in figure 11.3. These results show the ability to study nonlinear aerodynamic characteristics by the wing-flow method.

The balances and other equipment for these tests were largely designed by Harold I. Johnson, an engineer in the Stability and Control Section. A vacuum-operated automobile windshield-wiper motor was usually used to oscillate the model or the flap. A single 30-second dive covered the whole range of angle of attack and Mach number, providing as much data as might normally be obtained in a couple of weeks of wind-tunnel testing. As a result, the engineers were soon overwhelmed with data, but they worked

FIGURE 11.2. Drawing of semispan model of rectangular wing with full-span flap used in wing-flow tests.

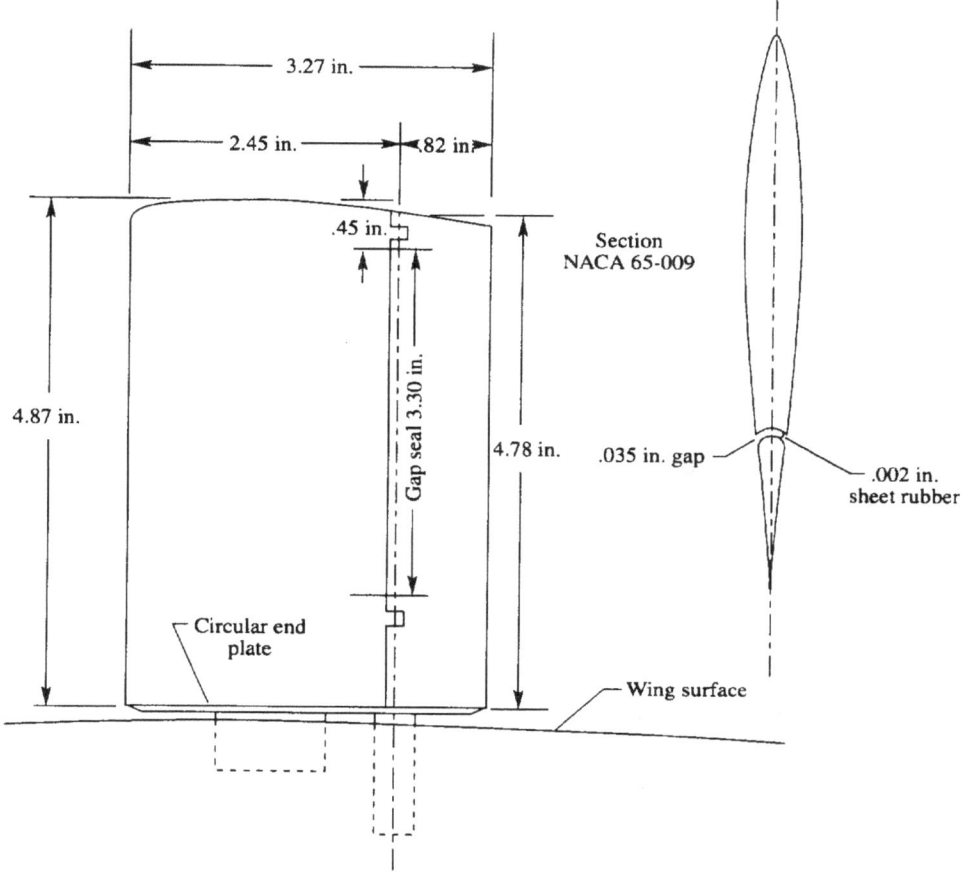

hard to evaluate it and to put out reports. As might be imagined, there was considerable criticism of the wing-flow method by some wind-tunnel specialists because of the small size of the models and the nonuniformity of the flow field. Nevertheless, the data obtained provided the only source of information at transonic speeds. Some vindication for the development of the method by the engineers in the Flight Research Division was later received when the wind-tunnel engineers installed a transonic bump on the floor of the Langley 7- by 10-Foot High-Speed Tunnel. The flow field over the bump was used to test small models by the same techniques as used for the wing-flow method.

In conventional wind-tunnel testing, the airspeed and model configuration are carefully set to the desired conditions before taking a reading on the balance. In an effort to obtain great accuracy, control surface and flap deflections are sometimes set by building separate inserts for each value of deflection. As a result the tunnel has to be shut down between tests at each control setting, which results in a very long time to complete the tests. To shorten the test time, the number of test points is reduced as much as possible. For example, I have frequently used wind-tunnel test data in which readings were taken at increments of five degrees in angle of attack and sideslip and in which control settings or flap settings were limited to a few values through the deflection range. One

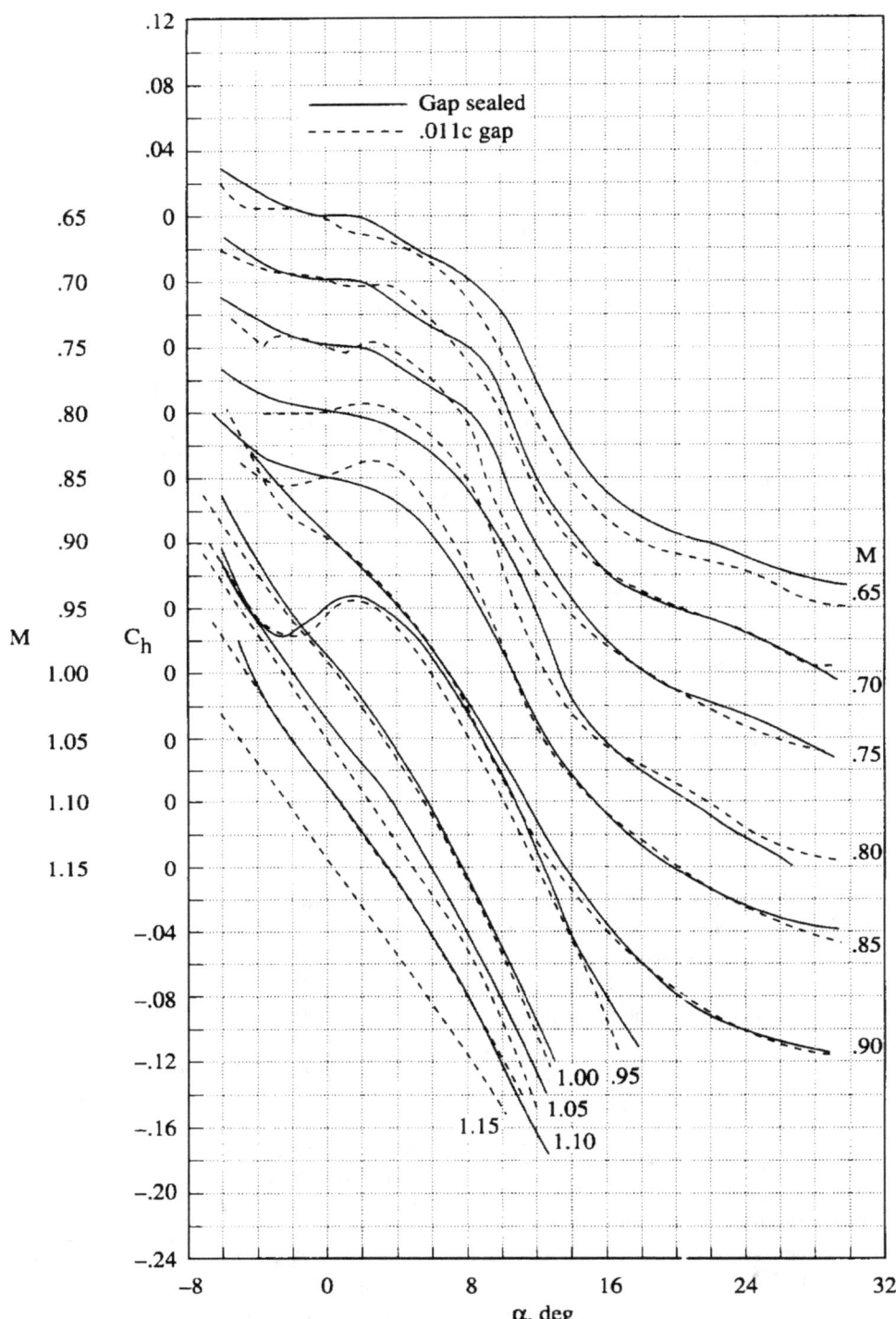

FIGURE 11.3. Typical data from wing-flow test. Plot of flap hinge moment as a function of angle of attack at various values of Mach number, gap sealed and unsealed.

(a) Mach number from 0.65 to 1.15 in steps of 0.05.

FIGURE 11.3. Concluded.

(b) Mach number from 0.9 to 1.0 in steps of approximately 0.01.

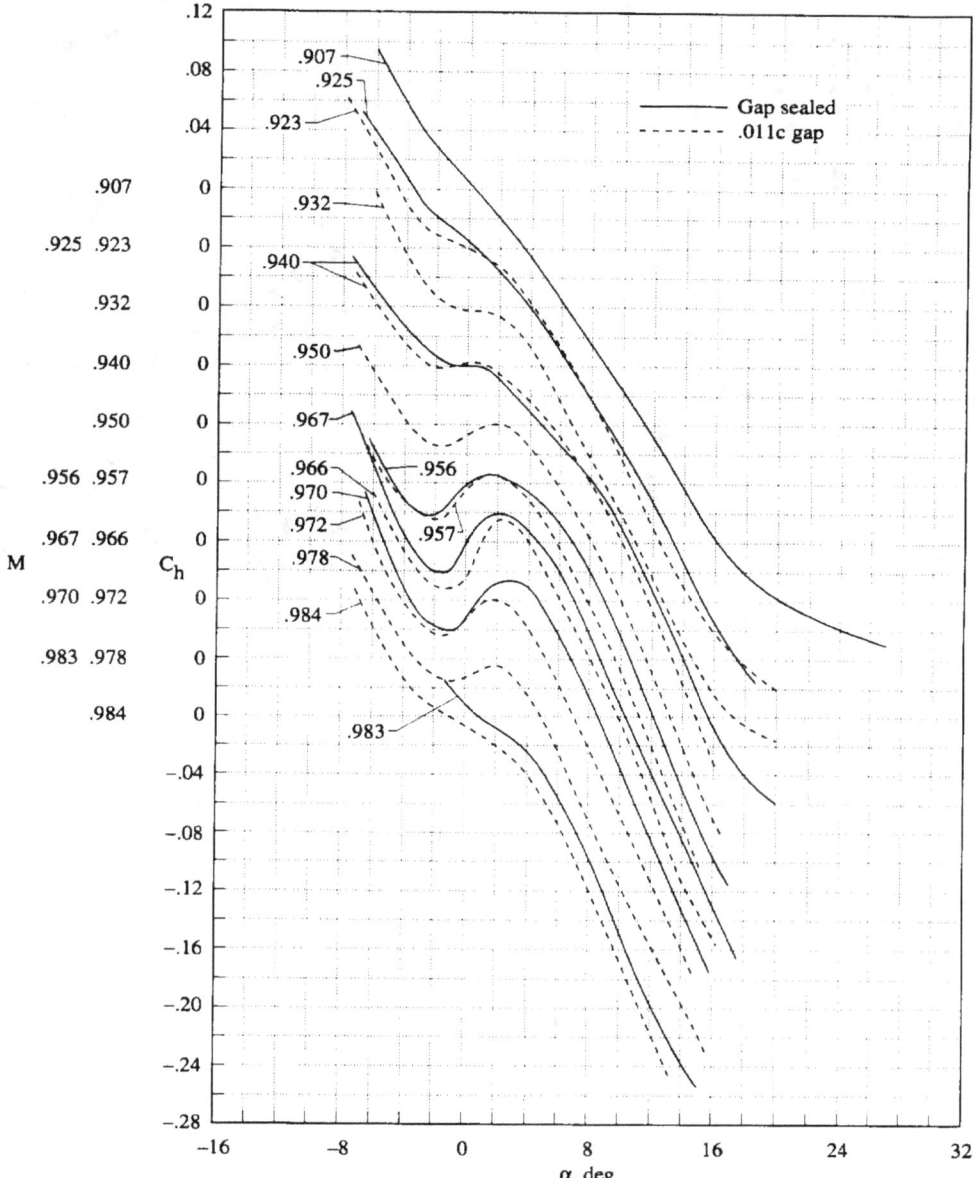

great advantage revealed by the methods developed for the wing-flow technique was that the data obtained were continuous through the range of variables. As a result, small irregularities or nonlinearities in the aerodynamic forces could be determined with as much accuracy as desired, whereas such features would be missed in conventional wind-tunnel testing. In flight at high speed, the variations of angle of attack and sideslip required to reach the limit loads on the airplane structure are frequently only a few degrees. Any small irregularities or nonlinearities in this range are therefore

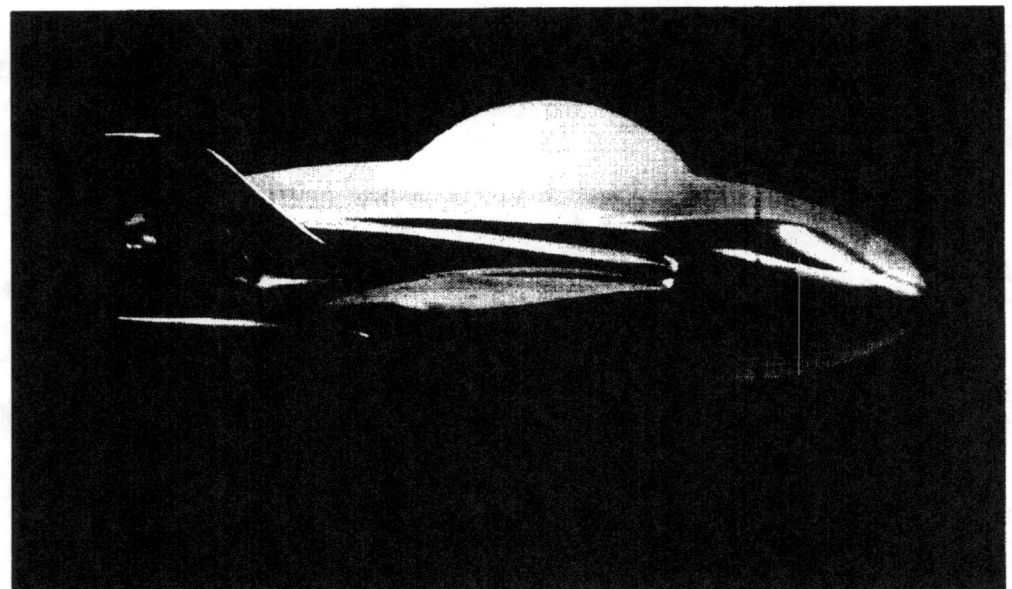

FIGURE 11.4. **Wing-flow model of Vought F7U-1 Cutlass. A half model is used, mounted on an endplate and curved to the contour of the P-51 wing section.**

of extreme importance. The effects of non-linearities in the variation of rudder hinge moment with angle of sideslip in causing snaking oscillations has already been discussed.

Another desirable feature of the wing-flow technique was that it allowed study of the effect of rate of change of variables as well as the effect of steady values. In actual flight, most flight maneuvers involve changing conditions, but the effects of the rates of change are not studied in conventional wind-tunnel tests.

As a result of these considerations, I and others have made suggestions to wind-tunnel engineers, from time to time, to incorporate provisions in their mounting systems and balances to allow tests under changing conditions. Some progress has been made in use of models with motor-driven control surfaces, but the desire to retain existing equipment and traditional techniques has generally prevented any change in the methods of mounting models or in the use of steady test conditions.

Because of the lack of any other source of transonic data, the results of the wing-flow tests were eagerly sought by the airplane manufacturers during the period following WW II. The tests of the 35-degree swept wing in figure 11.1 showed that the adverse stability and control characteristics at transonic speeds shown by most fighters with unswept wings were avoided by this model. As a result, the Chance Vought Company laid out the design of the F7U Cutlass Navy fighter as a tailless airplane with the same sweep and aspect ratio as the model. Some modifications were later made to incorporate the control surfaces, vertical tails, and fuselage, and a half model of the complete airplane configuration was tested (figure 11.4). The completed airplane is shown under test at the NACA Langley Flight Research Division in figure 11.5. This is probably the only case in history when the design of a new airplane was inspired by data taken in a 30-second test run.

The wing-flow testing continued for some years, but by 1950 Langley aerodynamicists developed the slotted-throat wind tunnel that enabled valid test results to be obtained

FIGURE 11.5. Full-scale Vought F7U-1 Cutlass.

through the transonic speed range. As a result, testing by the wing-flow and free-fall methods was discontinued.

Power Controls and Pilot-Induced Oscillations

With the development of airplanes capable of transonic and supersonic speeds, the use of manual control systems, even with the use of the accumulated knowledge of closely aerodynamically balanced controls and spring tabs, became impractical. Designers were forced to employ hydraulically actuated controls, even though the pilots and old-time designers were suspicious of the reliability of these devices.

Hydraulically actuated controls actually had a long background of experience, but not in the aeronautical industry. The first hydraulically actuated steering system for automobiles, later called power steering, was invented by a Harvard professor who lived in my home town of Belmont, Massachusetts. He installed the system in a 1923 Pierce Arrow. This system was developed without the aid of the theory of servomechanisms, which did not become available until the 1940's. This inventor did not succeed in convincing the automotive industry to use his invention until WW II, when it was installed in many Army trucks. After this demonstration of its practicality, power steering was soon adopted for automobiles and is now almost universally used.

Hydraulic controls were also used in large Navy gun turrets. These systems had excellent response and accuracy, but were much too heavy for aeronautical use.

The first hydraulic control systems to be installed in airplanes were called power boost controls, which means that a certain fraction of the force to operate the controls was fed back to the control stick so that the control forces would reflect the aerodynamic hinge moments on the control surfaces just as in a manual control system. The first installation of power boost controls was on the ailerons of a Lockheed P-38. The test pilot who made the initial tests of this system said it gave him the feeling of "supreme domination of the air." Evidently the ability to produce large values of rolling velocity in high-speed flight with light control forces was very desirable.

Installations of power boost controls on large airplanes, however, had less immediate success. These systems installed in the Douglas B-19, a large four-engine bomber, resulted in a violent pitching oscillation in the landing approach that almost resulted in a crash. Similar systems in the Martin Mars, a large four-engine flying boat, had many development problems.

The first fighter airplane to require power controls was probably the Vought F7U Cutlass. This airplane, a tailless design, required very wide chord elevons, which could not conveniently employ aerodynamic balance. To develop the hydraulic control systems for this airplane, the Vought company installed power controls in an F4U Corsair. This airplane was later made available to the NACA for research work.

This airplane, when flown by NACA test pilots, frequently encountered rather large-amplitude, unexpected, longitudinal oscillations when entering turns or pull ups. It was soon realized that these oscillations represented an instability of the combined system consisting of the human pilot, the airplane, and the control system. These oscillations were called pilot-induced oscillations, or PIO for short. After some practice, the pilots could fly without encountering these oscillations, and in fact, found it difficult to reproduce the condition for test purposes. They could, however, detect a tendency for the these conditions to exist. The hydraulic system could be turned off and the airplane flown with manual control, in which case there was no tendency for oscillations.

When the controls were operated on the ground with the hydraulic system turned on, everything felt perfectly normal, just like a stationary automobile with power steering. In this case, however, there was no way for the pilot to judge the response of the airplane. The closed loop involving the airplane, control system, and pilot was broken.

There was much interest in finding the cause of the pilot-induced oscillations and in determining a method to predict such problems on a new airplane before it was flown. I felt that the method used in analyzing the stability of closed-loop systems known as the frequency-response method could be used for this purpose. This method, by this time (about 1950), was available in textbooks. Also, a convenient graphical method called the Nyquist criterion could be used to predict the presence of stability or instability. Some new concepts were used in the application of this method, however. First, from handling-qualities experience. control forces rather than control displacements were believed to be the quantity of primary interest to the pilot because the control displacements on the elevator of a fighter airplane at high speed are very small. Second, the concept of considering the human pilot as a black box in a closed-loop system was practically unknown at that time, and reasonable estimates had to be made of a mathematical representation of the pilot's response.

The analysis was made using experimental data on the control system by oscillating the control stick and measuring the resulting control forces and elevator deflections. These tests were made at various values of frequency and amplitude. The normal acceleration of the airplane in response to the elevator motion was determined from well-established airplane stability theory, and the response of the pilot to airplane motion was estimated as a simple transfer function in which the pilot was assumed to apply a control force opposing pitching velocity with a constant time lag of 0.2 seconds. When these results were combined and the Nyquist criterion plotted, conditions of instability were clearly shown. The source of the instability was found to be friction in the hydraulic valve that admitted fluid to the servo cylinder. These valves had tight-fitting rubber seals to prevent leakage of hydraulic fluid and required a force of one to two pounds on the control stick to operate. Further study showed that this valve friction resulted in a

FIGURE 11.6. Simple simulator used to demonstrate to pilots the possibility of pilot-induced oscillations caused by characteristics of the power control system.

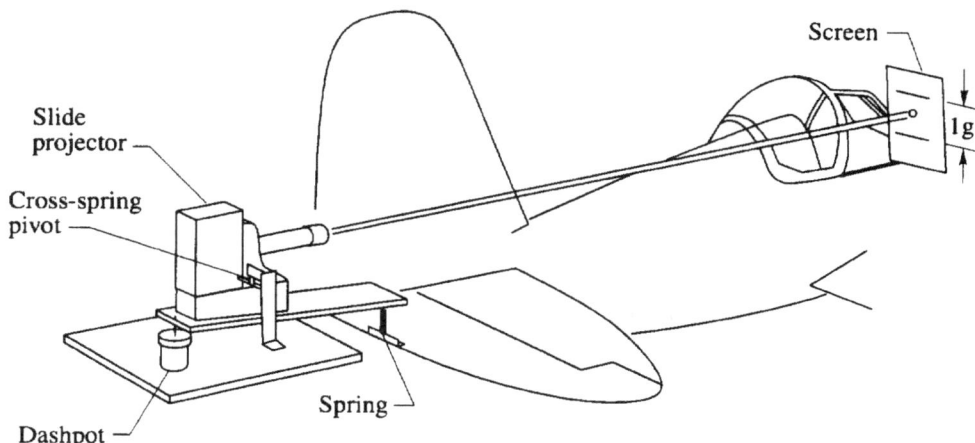

control force almost 180 degrees out of phase with the stick displacement at small amplitudes of motion.

I presented these results at a conference in Washington that was called by the Bureau of Aeronautics of the Navy Department to discuss power control systems. The other conferees were very interested in the data because most previous opinions of the problem had been based on speculation rather than on a rational analysis.

As previously mentioned, there was no indication of any control problem when the control stick was moved with the hydraulic system turned on and the airplane on the ground. A method was desired to demonstrate to the pilots that control difficulties might be encountered in flight. A very rudimentary simulator was built to provide this capability. As shown in figure 11.6, a slide projector was mounted on flexure pivots and driven by a spring attached to the trailing edge of the elevator. The motion was damped by a dashpot consisting of a piston with large clearance moving in a can full of oil. The damping and spring constant could be adjusted so that the response closely approximated the normal acceleration response of the airplane to elevator motion. The projector produced a spot of light on a screen alongside the cockpit with a motion of several

inches corresponding to 1 g change of acceleration. Two lines were drawn on the screen and the pilot attempted to move the light spot rapidly from one line to the other, simulating a 1g pull up. The response with the basic control system could be compared with that with the power control system operating. The response with the power control system on frequently showed an oscillation, which the pilots considered very similar to that obtained in flight.

In the design of modern airplanes, a mock-up of the control system and cockpit controls and instruments is often built to study possible control system problems. This type of simulator, known as an iron bird, can be very useful in studying control feel and response characteristics. Frequently, however, normal acceleration is displayed on a cockpit instrument that has low sensitivity in terms of pointer motion per g units. In flight, the pilot detects changes in acceleration through sensing the force acting on his body. Studies have shown that the sensitivity of the pilot to changes or oscillations in acceleration is very large. To study pilot-induced oscillations, therefore, an instrument with large sensitivity, such as the slide projector device used on the F4U, should be used to give the pilot sufficient stimulus to represent small changes in acceleration.

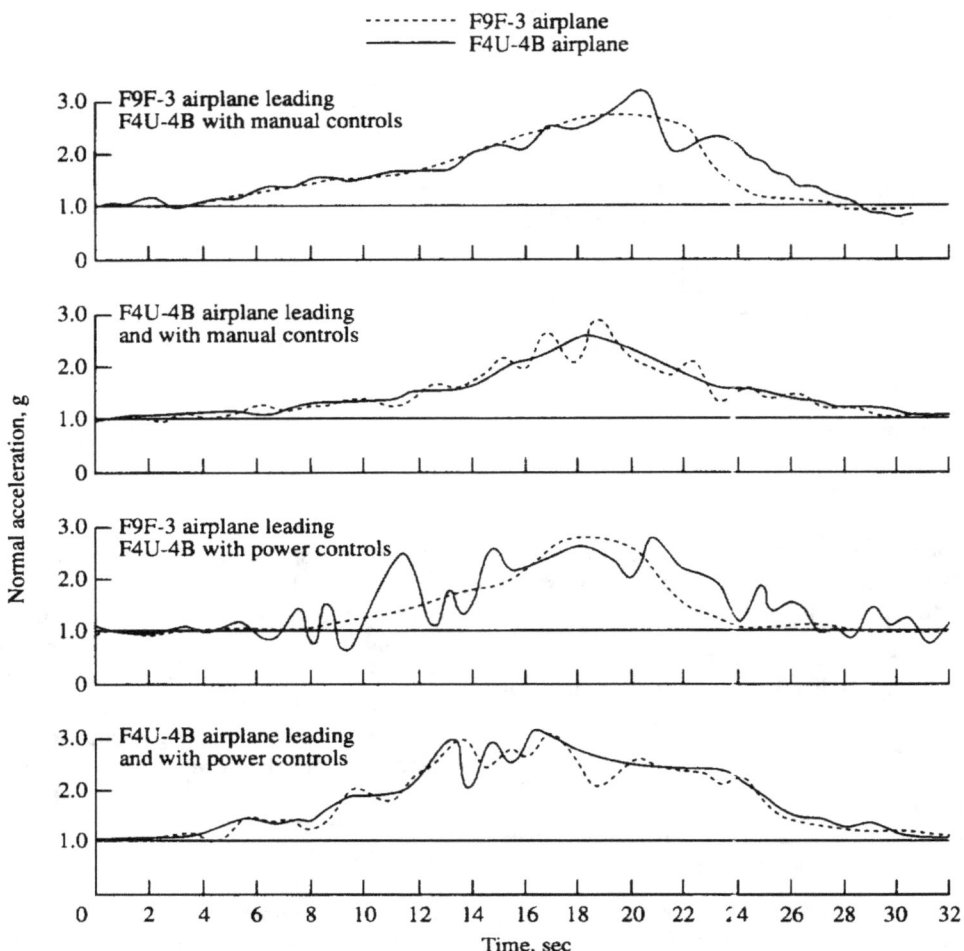

FIGURE 11.7. Normal acceleration during formation flights of F4U-4B airplane and F9F-3 airplane with alternate positions as lead airplane and following airplane; F4U-4B with and without power controls.

A test to compare the characteristics of different control systems in flight to determine their susceptibility to pilot-induced oscillations would also be desirable. One technique that was used in the F4U-4B tests is illustrated in figure 11.7. The F4U-4B was flown in close formation with an F9F-3 airplane, which had a manual control system with good control characteristics. First, the records of normal acceleration are shown with the F9F-3 in the lead and the F4U-4B with manual control. Second, the F4U with manual control is leading. In general, the records for the following airplane are expected to show some oscillations as the pilot applies control to maintain a close formation. Third, the F9F-3 is in the lead with the F4U-4B with power control. The large-amplitude oscillations of the F4U-4B as the pilot attempts to perform this tight control task are apparent. Finally, the F4U-4B with power control is in the lead. The F4U-4B flight path is somewhat unsteady. The F9F-3 is able to follow these oscillations with a well-damped and rapid type of response. These records show that a task requiring tight control is required to bring out the deficiencies in the control system.

Numerous studies of power controls and feel devices were made by the Flight Research

Division in the early 1950's (ref. 11.1). As aircraft manufacturers gained more experience with power control systems, many of the early problems disappeared as the importance of valve friction was recognized and means to keep the friction low were developed. Pilot-induced oscillations with other causes, however, remain a persistent source of trouble (ref. 11.2). Designers must try to take advantage of accumulated experience and modern analysis and simulation techniques to try to avoid these problems.

Human Response Characteristics

Autopilots for airplanes had been considered by inventors even before the first flight of the Wright brothers. For example, Sir Hiram Maxim in 1893 made a working model of a steam-powered gyroscope and servo cylinder to maintain the longitudinal attitude of an airplane such as his large steam-powered airplane. In 1913 Glenn Curtiss demonstrated hands-off flight with one of his biplanes controlled by an autopilot designed by Elmer Sperry. Later, in the 1930's, Sperry autopilots found practical use in transport airplanes. Anyone working with autopilots could not escape the realization that the human pilot, in controlling an airplane, worked on the same principle as an autopilot. That is, the pilot could sense the disturbances from a desired path and apply corrections through the control system to correct for the disturbances.

In the late 1930's and 1940's, theories became widely available to allow prediction of the controlled motion of an airplane when the action of the autopilot could be stated mathematically by means of a differential equation or by an expression known as a transfer function that related the input and output. With such techniques, the details of the mechanism could be neglected and the device replaced by a black box with the specified transfer function. It was realized that the human pilot could likewise be replaced

by a black box if a suitable transfer function for his behavior could be found. Such a formulation would allow prediction of the motion of an airplane under control of a human pilot and might be of value in predicting handling qualities.

Some early attempts at representing the response of a controlled airplane were made by Professor Otto Koppen and his students at MIT. The stability of a controlled airplane was first studied by Koppen in 1935 by assuming that the controls were moved in proportion to the angular deviation of the airplane from the desired attitude (ref. 11.3). Later this work was continued by two MIT students, Herbert K. Weiss (ref. 11.4) and Shih-Nge Lin (ref. 5.7), who took into account lag in the action of the controls. These reports gave very useful information on the action of closed-loop controls, but little attention was given to the question of whether the controls were moved by a human pilot or by an autopilot.

When I started work at Langley, Herbert K. Weiss was working with the Coast Artillery Board at Fort Monroe, also in the city of Hampton, Virginia. He was interested in human control of pointing of guns, or gun laying, as the British called it. We had discussions of the subject, and he described different methods of control application by a human pilot, such as control of position, rate, acceleration, and combinations of these quantities. These ideas were not immediately applicable to airplane control because the only mechanisms for performing these functions were heavy mechanical devices such as differentials and ball-disk integrators. These ideas, however, served to keep up my interest in the subject of human pilot characteristics.

An important advance in the analysis of the human pilot in controlling a dynamic system was made by A. Tustin, a British electrical engineer (ref. 11.5). His problem was to improve the tracking of a moving target by an electrically rotated turret in a tank and controlled by a human operator. He was able to characterize the human operator's response in terms of a simple control law involving a proportional and a lead term and a constant time lag caused by the combined effects of lags in the pilot's sensory and neuromuscular systems. In addition, he found that the human pilot response contained a superimposed random motion, which he called the remnant. He was able to use the pilot transfer function to determine modifications to the turret control law that would greatly reduce the tracking error. He also suggested methods for further analysis of human pilot response that have occupied researchers for many years since that time.

Inasmuch as military airplanes are used as gun platforms, the tracking accuracy is important in determining the probability of getting hits on the target. Tustin's success in improving the aiming accuracy of tank gun turrets gave hope that similar benefits could be achieved by applying human response theory to airplanes. Also, the handling-qualities requirements formulated as a result of the NACA research were all subjective measures, which were based on pilot opinion. The ability to use a mathematical representation of the human pilot gave hope that a more quantitative method of specifying the requirements might be possible. These applications gave an incentive for following closely the results of research in this field.

Early research was generally performed by recording the errors of the human operator in tracking a moving dot on a cathode ray tube and fitting the recorded response with a mathematical function that would match the performance of the human as closely as possible. Work of this kind was performed at MIT and was sponsored by the Wright Air Development Center at Wright-Patterson Field at several research organizations, including Franklin Institute, Systems Technology Incorporated, and the Cornell Aeronautical Laboratory. At Langley, the equipment for conducting such research was not readily available until the development of analog computers and simulators of various kinds. By this time the study of human response had become an extensive field of research with many research organizations

taking part. I assigned one of my engineers, James J. Adams, to follow this research and to conduct studies applicable to our interests. This work continued over a period of many years and extended into the period of the space program.

Landing Studies

During WW II, the wing loadings of airplanes increased greatly over the values used before the war, and high-lift flaps and large propellers resulted in a very steep glide. In landing these airplanes, pilots customarily kept the power on until just before touchdown to reduce the sinking speed. There was considerable speculation and study, however, on the problems that might be encountered if a power-off landing was required. One of the airplanes most extreme in terms of wing loading and sinking speed was the Martin B-26 bomber. A flight investigation was made on this airplane around 1944 to investigate piloting techniques required. In particular, runs were made in which the pilot glided the airplane toward the ground with successively increasing sinking speed to study the altitude at which the flare should be started and how accurately the pilot could judge the flare when it was started at an altitude much greater than had been used on most previous airplanes (ref. 11.6).

About 1947, I became interested in studying this problem analytically. Previous studies of the landing flare had assumed relatively small values of the glide angle. This assumption allowed the trigonometric terms involving the glide slope θ to be simplified by the assumptions $\sin \theta = \theta$, $\cos \theta = 1$. This assumption allowed calculation of the flare paths to be made by analytical methods. This assumption was not very accurate with the steep glide angles of the modern airplanes. I felt that flare paths could be calculated quite accurately without this simplification by use of the method of undetermined coefficients referred to previously. In this method, all terms in the equations are expressed as power series in the independent variable t

(time). Coefficients of like powers of t are collected, and the term involving each power of t is multiplied by an undetermined coefficient. In using the method, the lowest order coefficient may be determined from the initial conditions. Then the next higher order coefficient may be determined from this coefficient, and so on until all coefficients have been determined. In the work conducted, terms through the fifth order in t were used. Because the landing flare (in reverse time) is approximately a parabolic curve, the higher order terms rapidly decrease in size and the method is highly accurate.

I set up the method in tables (now called spreadsheets) to be filled in with calculated values by the female computers in the branch. They used mechanical Frieden or Marchant calculators in the work. Many landing flares were calculated for various values of the parameters. It is well-known that the shortest possible landing over an obstacle may be obtained by making a flare at maximum lift coefficient. The starting point of the flare, however, must be at precisely the right point to allow the wheels to touch the ground when the flight path is horizontal. My objective was to study flare paths that would allow a margin for error and to determine how much the landing distance would be increased as a result.

Despite much study, I was unable to determine a satisfactory criterion for the margin of error to be allowed. The error, of course, depends on the ability of the human pilot to judge altitude and sinking speed. Little work had been done on the subject at that time. Also, I came to realize that there were marked differences in the lift and drag characteristics of different airplanes that would affect the results. Airplanes like the B-26 with power off and flaps down would have high drag and a low value of lift-drag ratio throughout the landing maneuver, whereas delta-wing airplanes would have a decreasing lift-drag ratio throughout the flare. As a result, the data were never published. A report with a similar objective was published by Langley engineers Albert E. Von Doenhoff and George W. Jones, Jr. a few years later (ref. 11.7). They used numerical integration in calculating flare paths.

As an aside in connection with the paper by von Doenhoff and Jones, the reader is reminded that von Doenhoff worked in the Low Turbulence Tunnel Branch and was mainly noted for his airfoil studies and for a book by Ira Abbott and von Doenhoff on airfoils. It might be considered surprising that an engineer with this background would undertake a study of airplane landings. This versatility and broad range of interests, however, is typical of many of the leading engineers at Langley during this period. It is this versatility that allowed the engineers at Langley to take a leading role in the space program a few years later, even though this program involved disciplines very different from their former specialties. Von Doenhoff himself made an excellent analysis of space power systems comparing photovoltaic cells and thermoelectric generators (ref. 11.8).

Returning to the subject of landings, flare paths could be determined readily today by integrating the equations of motion on a digital computer. Also, if desired, simulated landings could be made by a human pilot in a simulator with realistic displays of the landing area. Such studies have no doubt been made in connection with the development of more recent airplanes.

One useful result that I did obtain from the study was to show that an airplane such as a delta-wing airplane, which has relatively high lift-drag ratio at low angles of attack and rapidly increasing drag as the angle of attack is increased, could probably be landed quite easily with power off by a human pilot. The flat glide in the approach allows starting the flare at a relatively low altitude. I gave a talk at a department meeting on this subject. This conclusion was confirmed, many years later, when an Air Force pilot safely landed an F-106, a delta-wing interceptor airplane, on the short runway (5000 feet) at Langley after his engine had failed at an altitude of several thousand feet.

CHAPTER 12

Measurement of Turbulence in the Atmosphere

The study of turbulence has been of interest to scientists and engineers for many years. Because of the random nature of turbulence, it is subject to the same types of analysis as random noise, such as is produced in electronic circuits by the random discharge of individual electrons or in a rarefied gas by the random collisions of molecules. The random motions of small particles in a liquid that result from bombardment of the particles by the molecules of the liquid is called Brownian motion. This phenomenon was analyzed by Albert Einstein in a famous paper written in 1905 (ref. 12.1). Turbulence became of interest to aerodynamicists first through studies of the boundary layer formed by air or liquid flowing over a solid boundary. The boundary layer could be either laminar or turbulent. The turbulent boundary layer was found to produce skin friction drag many times greater than that found in the laminar boundary layer. The tendency for the boundary layer to become turbulent depends strongly on the Reynolds number, a quantity that increases with the size of the model being tested, the speed, and the density of the test medium, and varies inversely with the viscosity of the medium. Early tests of small models in low-speed wind tunnels were found to give erroneous results because of the greater tendency of the boundary layer to remain laminar on the model as compared to that on a full-scale airplane. The boundary layer was studied extensively by Hugh L.

Dryden, later a director of the NACA and by a famous British scientist, G. I. Taylor, as well as many other noted scientists (ref. 12.2). I first became aware of these studies when I wrote my bachelor's thesis on boundary layers in 1939. Soon the subject of turbulence became a discipline of interest in many fields and could easily become the focus of a lifetime career.

The turbulence in the atmosphere is a random three-dimensional turbulence field that can extend for large distances compared to the size of an airplane penetrating it. Early applications of turbulence theory to atmospheric turbulence were made by T. von Kármán in 1934 (ref. 12.3) and by G. I. Taylor in 1935. These reports were written before I started work, but I made an effort to review them after I had worked at Langley for some years.

When an airplane flies through turbulence, large loads are imposed on the structure that may be critical for the design strength of the airplane or repeated smaller loads may lead to fatigue failure of structural components. Also, the violent motions of the airplane may cause airsickness or fatigue of the crew and passengers. As a result of these effects, atmospheric turbulence was recognized as an important subject for research.

The type of turbulence most frequently studied in the theoretical analyses is called homogeneous isotropic turbulence. Homogeneous means that the turbulence is the same at any location, and isotropic means that the turbulence field is the same in any direction. In such turbulence, an airplane would experience the same type of disturbances flying in any direction. Not all atmospheric turbulence has these characteristics, but this theory appears to be useful in many practical applications. Two terms associated with this turbulence are the autocorrelation function and the power spectrum. The autocorrelation function shows how, on the average, the turbulent velocity at some point is related to the turbulent velocity at some other point. Thus, if a gust hits an airplane, experience has shown that the velocity is approximately the same at the wing tips as at the centerline. The velocity at some point far away from the airplane would, on the average, be unrelated to the velocity at the airplane, which shows that the correlation function approaches zero at large distances. The power spectrum considers the turbulence in the atmosphere to be composed of superimposed sinusoidal waves of different wavelengths. The spectrum shows how the amplitude of a wave, on the average, varies with the wavelength. It can be shown mathematically that the correlation function is related to the power spectrum, so that if one is known, the other can be calculated.

The shape of the spectrum of vertical gusts as predicted by von Kármán on a plot of the gust power (that is, the square of the gust velocity in each increment of frequency) as a function of frequency shows that the power is constant at very low frequencies. Then at some frequency, the power starts to decrease, and falls off as the frequency to the −5/3 power. This region of the spectrum is called the inertial subrange, indicating that the mass effects are important. This range is usually most important in influencing the loads and motions of an airplane. At still higher frequencies, the spectrum was shown by W. Heisenberg, the famous physicist, to fall off very rapidly and vary as the frequency to the −7 power (ref. 12.4). He called

this part of the spectrum the viscous subrange, indicating that viscosity effects in the air are important. When I became aware of Heisenberg's report, I had it translated and published as a NACA Technical Memorandum. I thought that this part of the spectrum might be of interest for airplanes, but it turns out that the wavelength is so short in this region that it has a negligible effect on full-scale airplanes, though it may be of some interest in wind-tunnel studies of turbulence.

The work on atmospheric turbulence made it apparent that the accuracy of turbulence measurements made in flight would be important and that studies to show how well the assumption of homogeneous, isotropic turbulence would apply in flight would be of interest. As a result, studies made of these subjects are presented in the following sections.

Analytical Studies of the Accuracy of Turbulence Data from Flight Records

In the time period of the early 1950's, numerous applications occurred in flight research for the analysis of random data. Records requiring this type of analysis were obtained, for example, in studies of pilot tracking error, in the response of airplanes to turbulence, and in records of human response to arbitrary inputs. The approach used was to analyze the record by fitting it with a Fourier series, which showed the amplitude of the harmonics at various frequencies. The earliest attempts at this type of analysis used a Coradi rolling sphere analyzer, a delicate instrument that mechanically read the amplitude of a component at a given frequency when the record was carefully traced by manually moving a stylus along the record. Later, when electronic computers became available, the records were digitized on punch cards and power spectral data were obtained, but the process was quite lengthy

FIGURE 12.1. Error in harmonic amplitude due to finite length of record for various phase angles of harmonic.

(a) Examples of waves that just fit within a rectangular band (top).

(b) Plot showing ratio of error to line width as a function of ratio of record length to wavelength (bottom).

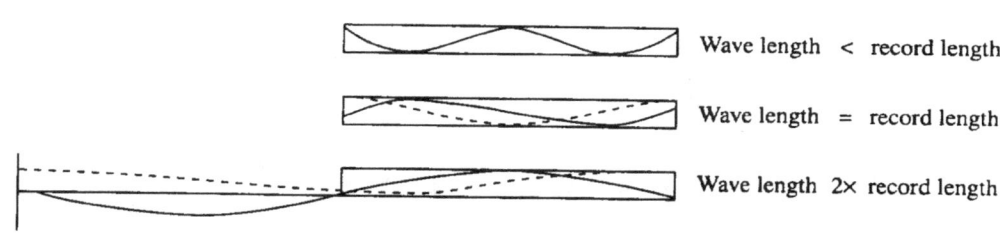

Wave length < record length

Wave length = record length

Wave length 2× record length

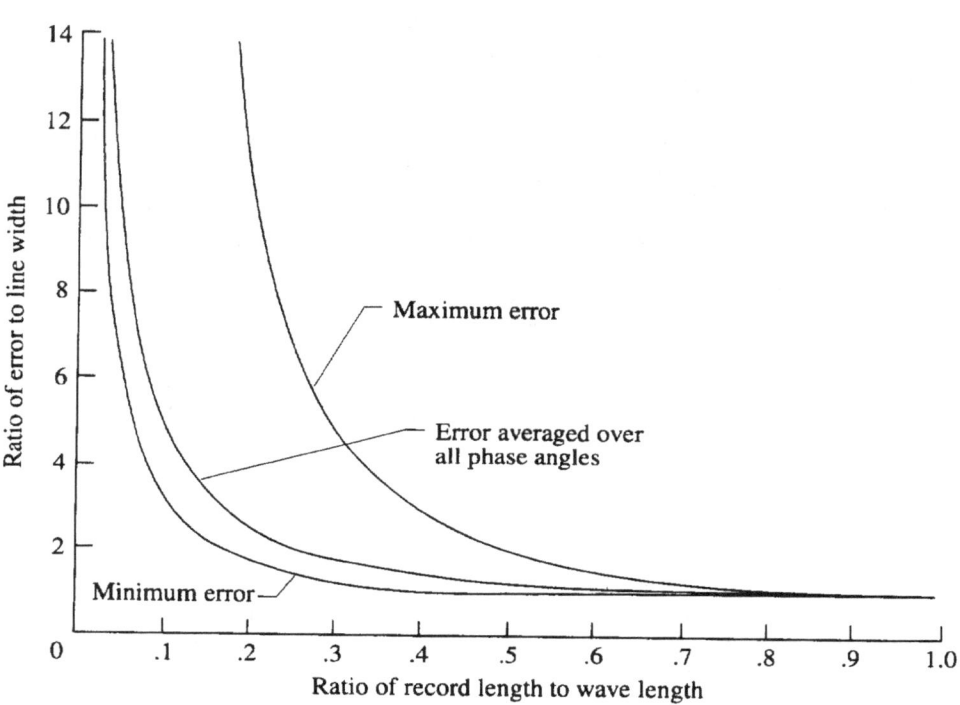

with the capabilities of computers available at that time. Now, of course, the availability of data recording in digitized form, the use of the fast Fourier transform, and the high speed of computation makes the analysis of such data a routine procedure.

One question that came up, particularly in the analysis of atmospheric turbulence, was how long a record was required to obtain sufficiently accurate data on the low-frequency components of turbulence. In the case of the early studies, a problem apparent to the researcher was that the time history record obtained as a trace on oscillograph film had a certain width. Frequently, engine vibration picked up by the instrument made this trace considerably wider than that of the trace under smooth conditions. The width of the trace could be observed on a physical record or could be considered as indicative of an error band that depended on the accuracy of the instrumentation involved. The actual value of the recorded quantity at any time could be anywhere within the width of the error band. A study was made to determine how large a variation in the amplitude and phase of a given harmonic component could occur without going outside the boundaries of the trace over the length of the recorded data. In particular, it should be noted that attempts were made to obtain data on harmonics with wavelengths considerably

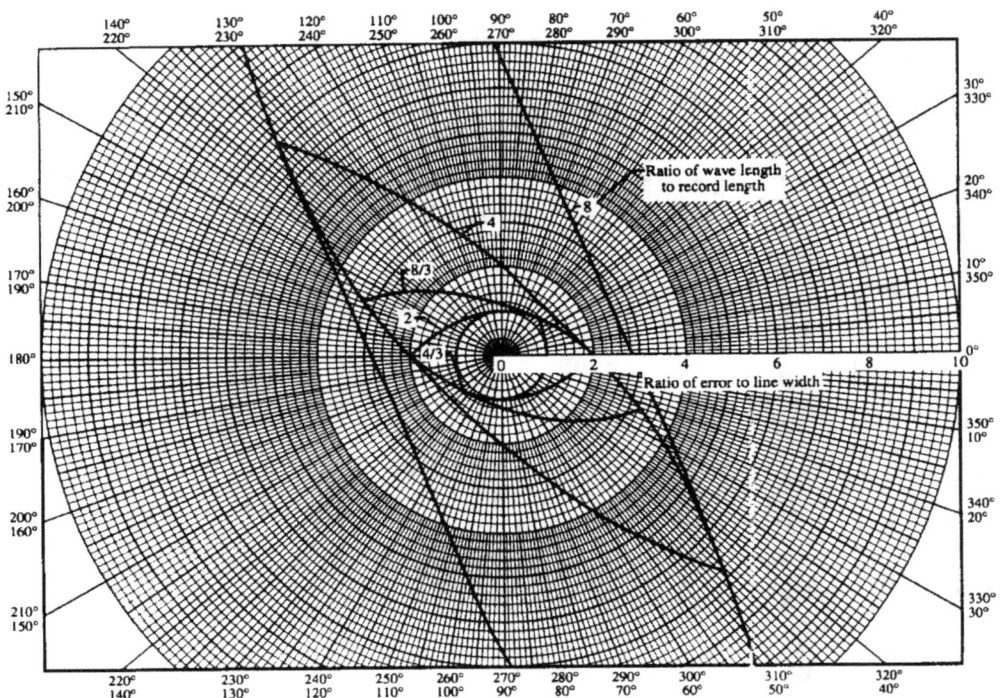

FIGURE 12.2. Plot of amplitude versus phase angle for waves with various ratios of wavelength to record length.

longer than the length of the time history available. This analysis involved relatively straightforward trigonometric manipulations. One reason for showing the results here is that the resulting plot of amplitude versus phase angle of the harmonics at various wavelengths has an unusually artistic form.

First, in figure 12.1(a) are shown several examples of harmonic components of waves that just fit within a rectangle with length equal to the length of the record and width equal to the error band. In practice, the record is many times longer than its width, but the figure shows a much shorter and wider rectangle for clarity. As shown, so long as the wavelength is less than the record length, the maximum error of the component is the same as the width of the error band. When the wavelength of the harmonic component is larger than the record length, however, the error of the component exceeds the width of the error band by an amount depending on the phase angle of the component. Figure 12.1(b) shows the ratio of error

to line width as a function of ratio of record length to wavelength for different conditions of phase angle of the harmonic. The conditions shown are those giving the maximum error, those giving the minimum error, and the error obtained by averaging over all phase angles. Inasmuch as the phase angles in power spectral analysis are assumed to be uniformly distributed, the latter curve is probably more representative of what could be obtained with a statistical analysis.

Figure 12.2 shows the polar plot of amplitude versus phase angle for harmonics of different ratios of wave length to record length. The nesting of the various lenticular curves presents an unusual artistic effect not often seen in theoretical plots.

This analysis was never published, mainly because more sophisticated approaches were being developed by mathematicians with much more knowledge of statistical analyses than I had. The next section describes a flight investigation to make accurate measurements

FIGURE 12.3. Photograph of F9F-3 airplane showing special instrumentation for turbulence measurements.

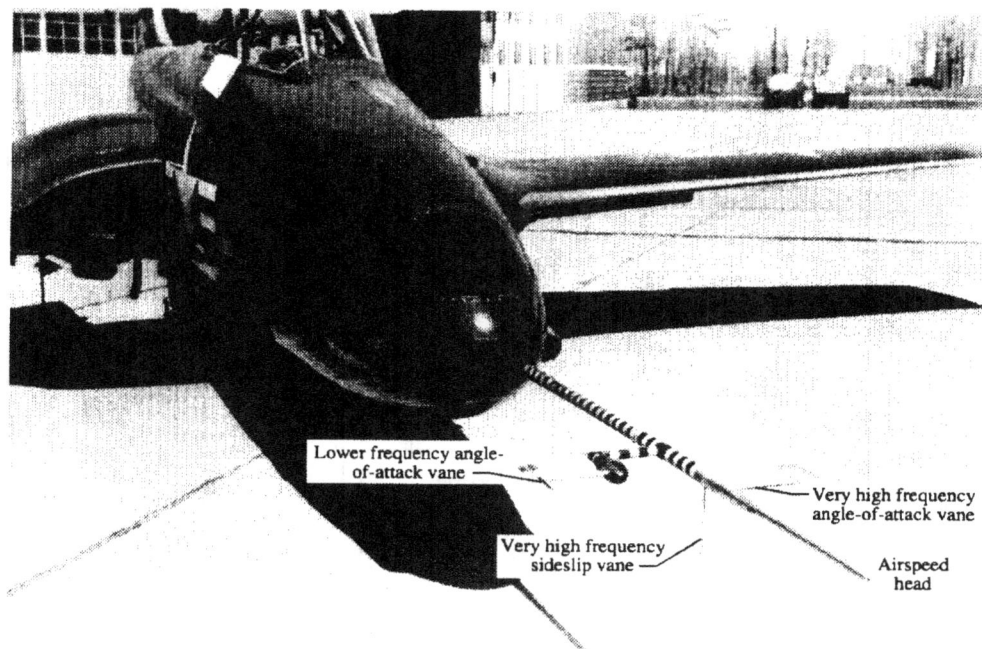

of the spectrum of atmospheric turbulence that uses the most advanced statistical techniques available at the time.

Measurements of the Spectrum of Atmospheric Turbulence Over a Large Range of Wavelengths

A much more important source of error than the error introduced by the instrumentation is the variability of the power spectral data caused by the random nature of the process being analyzed. Thus, various samples of a given process taken at different times will show different harmonic content, particularly at the longer wavelengths. To obtain an accurate measure of the power spectrum, many records should be analyzed and the results averaged or one very long record taken under uniform conditions should be analyzed. As a

result of these considerations, a rule of thumb is that approximately ten to twenty wavelengths of the lowest frequency under consideration should be contained in the record. Obviously, in considering atmospheric turbulence that contains much energy at wavelengths of several thousand feet, a very long record in uniform conditions of turbulence is required.

In my position as head of the Stability and Control Branch of the Flight Research Division in the early 1950's, I was able to plan programs of flight research and had available test airplanes and instrumentation to perform research studies. One of my engineers, Robert G. Chilton, a recent MIT graduate, was familiar with the latest mathematical techniques to analyze records of turbulence. The measurement of atmospheric turbulence was officially assigned to the Aircraft Loads Division, but the results of such studies were of immediate use in my branch to study response of airplanes to turbulence. Inasmuch as accurate measurements over a wide range of wavelengths

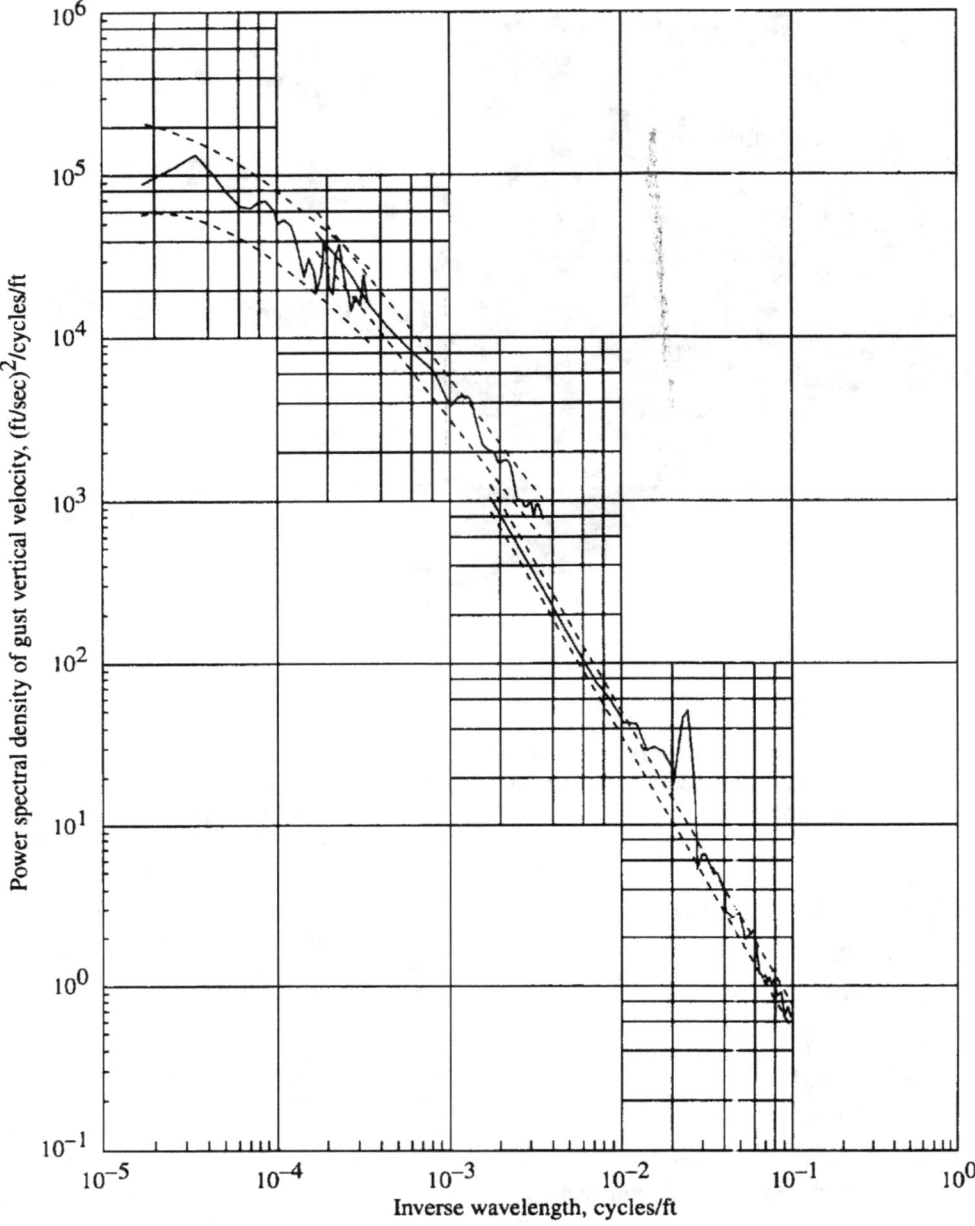

FIGURE 12.4. Power spectrum of gust vertical velocity for wavelengths of 10 feet to 60,000 feet. Dotted lines show 95 percent confidence bands.

were not then available, I obtained approval to conduct a flight test to obtain these data. These tests used instrumentation with greater accuracy than had been used previously, such as a sun camera (rather than a gyroscope) to determine airplane pitch angle and a high-frequency vane to measure angle of attack.

The airplane and some of the instrumentation are shown in figure 12.3.

With the aid of this equipment, a flight was made at an altitude of 1700 feet over a distance of 170 miles between Norfolk and Baltimore with a Navy F9F-3 Cougar fighter airplane. The weather conditions were favorable in that a cold front had just passed through the entire area, which gave a strong ground wind and consistent clear-air turbulence over the entire path. The results of this test are presented in a NACA Technical Note (ref. 12.5).

The measured spectrum of turbulence is shown in figure 12.4. To obtain maximum accuracy in different frequency ranges, different portions of the record were analyzed with different spacing of the recorded points. Thus, for high frequencies, a shorter portion of the record with more closely spaced points was analyzed. As a result, the plotted curve is broken into sections. Since these tests were made, many studies have been made by NASA, the Air Force, and other organizations to obtain turbulence measurements at various altitudes and meteorological conditions. It is believed, however, that the study presented in reference 12.5 is still one of the most useful for comparing the spectrum of atmospheric turbulence with the theoretical predictions over a wide range of wavelengths.

It is evident that my early analysis has little relation to the more important problems of obtaining accurate power spectral data. The artistic results of the analysis, therefore, may be considered more a mathematical oddity than a useful analytical result. Subsequent developments of turbulence theory by experts in this field, however, were followed closely and allowed useful data to be obtained on the spectrum of turbulence or in other studies involving random data.

CHAPTER 13

Gust Alleviation

The response of airplanes to gusts has been a subject of concern to airplane designers since the earliest days of aviation. The Wright Brothers, flying in high winds at low altitude on the seashore at Kill Devil Hills, North Carolina, purposely designed their gliders and their powered airplane with low or negative dihedral to avoid lateral upsets due to side gusts. Pilots soon found the necessity of seat belts to avoid being tossed out of the seats of their machines when flying through turbulence. One of the first women pilots, Harriet Quimby, and her passenger, flying without seat belts in a Bleriot airplane over Boston, Massachusetts, in 1911, were unfortunately thrown out of their machine and fell 1500 feet to their deaths. In more recent years, gusts have been recognized as one of the sources of critical design loads on airplanes, as well as a source of fatigue loads due to repeated small loads. In addition to these design problems, many people are susceptible to airsickness when flying through rough air.

Despite the long interest of airplane designers in the effects of turbulence, very few efforts have been made to design airplanes with reduced response to gusts. Even in 1995, none of the commonly used transport airplanes or general aviation aircraft were equipped with gust-alleviation systems.

In the past, several attempts have been made by airplane designers to build air-planes with reduced response to turbulence. All of these attempts were characterized by an intuitive approach with no attempt at analysis prior to flight tests, and all were notably unsuccessful.

One of these airplanes (figure 13.1) was designed by Waldo Waterman. It had wings attached to the fuselage with skewed hinges and restrained by pneumatic struts that acted as springs. The effect of the skewed hinge was to reduce the angle of attack of the wing panels when they deflected upward, and vice versa. The response to gusts was not noticeably reduced from that of the airplane with the wings locked, probably because the dynamic response of the system was not suitable. Also, the degree of flexibility of the wings was limited because deflection of the ailerons would deflect the wings to oppose the aileron rolling moment, which resulted in reduced or reversed roll response.

The effect of wings with skewed, spring-loaded hinges is similar to the effect of bending of a swept wing. Airplanes with swept wings do have smoother rides in certain frequency ranges and suffer from reduced aileron reversal speed when compared with airplanes with unswept wings.

A similar method that has been tried in flight is to incorporate springs in the struts of a conventional strut-braced high-wing mono-plane. This method may be likened to the springs used in an automobile chassis to

reduce bumps. This method has also proved ineffective, probably because of the slow dynamic response of the system.

Other schemes involving wing motion have been proposed from time to time, and some of them have been investigated in wind-tunnel tests or in flight. One method that has been given considerable attention is the "free wing" concept. In this method, the wing is pivoted with respect to the fuselage about a spanwise hinge ahead of its aerodynamic center, and its angle of attack is controlled by a flap on the trailing edge. A serious disadvantage of this method is that up-flap deflection must be used to trim the wing at a high-lift coefficient for landing. This flap obviously reduces the maximum lift, just the opposite from what is normally obtained with a downward-deflected landing flap. Also, the dynamic response of the wing may be too slow to provide reduction of the accelerations due to high-frequency gusts.

In England, shortly after WW II, a large commercial airplane called the Brabazon was designed. In the design stage, a system was incorporated to reduce wing bending due to gusts by operating the ailerons symmetrically to oppose bending due to gusts. The ailerons were to be operated by a mechanical linkage connected to the wing in a way to be moved by wing bending. The system was abandoned before the airplane was flown, and the airplane never went into production. Nevertheless, the project stimu-

lated interest in a flight project at the Royal Aircraft Establishment (RAE) in which a system of this type was tried in a Lancaster bomber. This system used a vane ahead of the nose as a gust detector to operate the ailerons symmetrically through a hydraulic servomechanism. The system was built with little preliminary analysis, and when the pilot engaged the system in flight for the first time, the flight in rough air seemed noticeably more bumpy than without the system. By reversing the sign of the gain constant relating aileron deflection to vane deflection, the ride was made somewhat smoother. Later, an analysis by an RAE engineer named J. Zbrozek showed the reasons for the unexpected behavior. These reasons will be mentioned later in the presentation.

Another experimental program was conducted on a C-47 airplane by the Air Force. This system was similar to that originally planned for the Brabazon. The ailerons were arranged to deflect symmetrically upward with upward wing bending, and vice versa, by means of a linkage which added a component of this deflection to that of the conventional aileron linkage. Since the wing deflection provided a large driving force, no servomechanism was required, and as a result, a system of high reliability was expected. The system suffered from the same objections as the one tested on the Lancaster. In addition, the inertia of the ailerons combined with flexibility of the operating linkage caused the aileron deflection to lag behind the wing deflection. Such a system is very conducive to flutter. To avoid flutter, the ratio between the aileron deflection and wing bending had to be kept to a very low value. As a result, the system was unable to provide more than 9 percent reduction in wing bending moments, a rather small improvement.

Despite the discouraging results of these experiments, the advantages of gust alleviation remained worthwhile. As a result, a project was initiated at Langley to study this subject. These activities are described in the following section.

Background and Analysis of Gust Alleviation

I became head of the Stability and Control Section of the Flight Research Division in 1943. During the wartime years, there was little difficulty in deciding on the type of work to be done by the section. Most of the work was concerned with flying qualities or with improving the control systems of airplanes. Some of this work has been described in the preceding sections or may be further seen from my list of reports (appendix II). By 1947, jet airplanes with power control systems were being developed, and the use of automatic control to improve the stability characteristics of airplanes was a rapidly developing field. Applications of automatic control were therefore important subjects of research. The change in emphasis was recognized in 1952 when I was made head of the Guidance and Control Branch of the Flight Research Division.

In the military services, the new technology of automatic control was largely applied in the development of guided missiles. At Langley, the Pilotless Aircraft Research Division, under Gilruth, had been established to use missiles for aerodynamic research. Some of the engineers in that division, however, became interested in guided missiles. Gilruth and William N. Gardiner started to perform tests on a missile with infrared guidance that used gyroscopic stabilization and employed rocket propulsion techniques developed for aerodynamic testing. It later turned out that this missile was very similar in concept to the Sidewinder missile developed by the Navy, which was one of the first air-to-air guided missiles widely employed by the armed services.

The Deputy Director of the NACA at that time was Dr. Hugh L. Dryden, a noted scientist, who was also a lay preacher in the Methodist Church and a religious man. He did not like to see the NACA centers unnecessarily involved in military work. He issued a directive that no work with military applications should be done at the NACA centers unless requested by the military services.

From the standpoint of a stability and control engineer, the study of missile guidance systems was one of the most interesting and challenging fields available at that time. Nevertheless, I had the same feelings as Dr. Dryden concerning the desire to avoid emphasis on military projects. With this area of research ruled out, I was faced with the question of what type of automatic control research with peacetime applications was of most interest. At that time, airplanes with swept wings were starting to be fitted with yaw dampers to improve the damping of lateral oscillations, particularly at high altitudes. Yaw dampers, however, did not require much research. In most cases, companies were able to hook up a rate gyroscope through an analog amplifier to the power control actuator on the rudder, and the system worked very well in damping out lateral oscillations.

After much thought, I concluded that one field in which automatic control could be applied and which had not previously been studied to any great extent, was gust alleviation. A gust-alleviation system is one that provides the airplane with a smooth ride through rough air. At that time, all transport airplanes had piston engines and flew at altitudes below the top of storm clouds. Fear of airsickness was a common problem and was a deterrent to many people who wanted to travel on commercial airlines.

I was already familiar with one study of gust alleviation that had been made. Philip Donely, who was then a branch head in the Loads Division, had formerly been designer of the gust tunnel at Langley. In this wind tunnel, a model was catapulted at flying speed through a vertical jet of air to simulate a gust in the atmosphere. Evidently while doing this work, Donely had come across a report by René Hirsch in France on a study of gust alleviation done as a doctoral thesis and published in 1938 (ref. 13.1). Donely brought this report to my attention. Hirsch

had devised a gust-alleviation system in which the halves of the horizontal tail were attached by chordwise hinges. These surfaces were connected by pushrods to flaps on the wing. On encountering an upward gust, the tail halves would deflect up, moving the flaps up and thereby offsetting the effect of the gust. Other features of the system made the airplane insensitive to horizontal gusts and to rolling gusts, all without adversely affecting the ability of the pilot to control the airplane. These many ingenious features are too complex to discuss herein.

I was very impressed by the report for two reasons. First, Hirsch had performed an analysis to determine the relations between the tail, elevator, and flap hinge-moment characteristics; linkage ratios; and other parameters so that the flaps moved exactly the right amount to offset the effect of a gust. This analysis required consideration of so-called stability derivatives for all these components. At that time, stability derivatives were used to describe the stability characteristics of an airplane, such as, the variation of pitching moment with angle of attack C_{m_α} and the variation of lift coefficient with angle of attack C_{L_α}. Prior to 1938, very few people had ever considered similar quantities for control surfaces or flaps, such as variation of flap hinge moment with angle of attack or variation of elevator hinge moment with deflection. Hirsch's analysis required a multitude of these quantities, as well as linkage ratios and other quantities describing the mechanism, all tied together by algebraic equations that were solved to obtain the desired design characteristics of the system.

The second important contribution by Hirsch was that he tested his system with a dynamic model. I have previously mentioned some of my experiments with a dynamic model. In Hirsch's case, the model was mounted in a wind tunnel so that it was free to pitch and to slide up and down on a vertical rod. The tunnel had an open throat and was equipped with a series of slats ahead of the test section, similar to a venetian blind, that could be deflected to produce an abrupt change in the flow direction. When Hirsch tested his model in this artificial gust with the system locked, it immediately banged up against its stop at the top of the rod. With the system working, however, it showed only a small disturbance and settled back to stable flight.

Hirsch was so impressed by these results that he devoted most of his professional life to demonstrating the system in flight. I later corresponded with Hirsch, visited him in France, and saw his airplanes. This part of the story will be mentioned later. At the time I read the thesis, however, I simply realized that gust alleviation was a feasible idea and that with automatic controls, it might be possible to do the job more simply than Hirsch had done with his complex aeromechanical system.

I started, in 1948, to make analyses of the response of example airplanes to sinusoidal gusts and the control motions that would be required to reduce (alleviate) the response of the airplane or the accelerations that would be applied to the passengers. The method used was what has been described earlier as the frequency-response method, which had the advantage that the calculations were greatly simplified when compared with calculating response to discrete gust inputs. These early studies showed, as was known from experience, that control by the elevators alone was ineffective. Control by flaps on the wing was thought to be promising, but the analysis showed that in many cases, the flaps would produce excessive pitching response of the airplane. In general, control by a combination of flap and elevators was required. With flaps alone, successful results could be obtained only by careful attention to the pitching moments applied to the airplane by the gust and by the flaps.

A survey was also made of previous attempts at gust alleviation on full-scale airplanes. These included an airplane made by Waterman with wings pivoted on skewed hinges, a DC-3 modified by the Air Force so that the ailerons deflected symmetrically in response to wing bending, and an Avro Lancaster bomber modified by the British in

which the ailerons were moved symmetrically in response to gusts sensed by a vane on the nose. All these attempts, made without any theoretical analysis, had been unsuccessful in that very little alleviation of airplane accelerations was obtained. My analysis showed why each of the attempts had failed. In general, the lack of attention to pitching moments produced by the control surfaces was responsible. In the case of the Lancaster, for example, the symmetric deflection of the ailerons in the upward direction proportional to a positive change in angle of attack produced a positive, or upward, pitching moment. A positive variation of pitching moment with angle of attack represents a decrease in longitudinal stability. This decrease in stability increased the response to low-frequency gusts, which resulated in an increase in the bumpy ride experienced by the pilots. In addition, the aileron motion reduced the damping of the wing bending oscillation, which increased the effect of structural oscillations on the sensation of the pilots. In considering the effect of the ailerons on the pitching moments, it should be realized that a symmetric upward deflection of the ailerons produces a direct effect on the pitching moments of the wing. It also changes the wing lift distribution to produce an increased downwash on the tail due to up aileron deflection, which further increases the destabilizing variation of pitching moment with angle of attack. My analysis showed that the downwash effects on the tail due to the deflection of flaps or ailerons on the wing are very important in designing a gust-alleviation system.

After making these preliminary studies, I analyzed a system in which a gust-sensing vane mounted on a boom ahead of the nose was used to operate flaps on the wing through a hydraulic servomechanism. Any gust-alleviation system working on this principle reduces the lift produced by a change in angle of attack. For complete alleviation, the lift due to angle of attack is reduced to zero. Since the pilot maneuvers the airplane by changing its angle of attack, this system would prevent the pilot from making any longitudinal maneuvers. To restore this capability, the input from the control stick, normally used to move the elevators, was also fed to the flap servomechanism. With this arrangement, when the pilot moves the stick back to make a pull up, the flaps first go down to produce upward lift. Then, as the angle of attack increases in response to the elevator motion, the flaps move back to neutral and the pull up is continued with the airplane at a higher angle of attack. The result is a faster response to control motion than obtained with a conventional airplane. This type of control, in later years, has been called "direct lift control" and is advantageous for control situations requiring rapid response.

The analysis also showed that provision had to be made for avoiding excessive pitching moments due to the gusts and flap deflection. Since the flaps were moved in proportion to angle of attack, any pitching moment from the flaps contributed directly to the variation of pitching moment with angle of attack, which determines the longitudinal stability of the airplane. For satisfactory stability, the pitching moment due to angle of attack must be kept within prescribed limits. The additional contribution due to the gust-alleviation system had the possibility of greatly exceeding these limits, which made the airplane either violently unstable or excessively stable. It was possible to solve the equations to determine the flap and elevator motion to completely offset both the lift and pitching moments applied to the airplane. This analysis showed that this objective could be attained in two ways. In one method, both the flap and elevator had to be moved in response to the gust, but the elevator motion was not in phase with the flap motion and generally had to lag behind the flap motion. In the other method, the elevator and flap moved in phase, but the downwash from the flaps on the tail had to be of opposite sign from that normally encountered. That is, down flap deflection had to produce an upwash at the horizontal tail.

Having reached this stage in the analysis, the results appeared sufficiently promising for a report on the results and a flight program to demonstrate a gust-alleviation system. A job order request was submitted about June 1948 to obtain official approval for this work. This job order is presented here to illustrate the type of request required to get a research program approved.

Title: Theoretical and Experimental Study of Means to Increase the Smoothness of Flight through Rough Air.

Est. Man Hrs: 3000   Cost: $6000

Description: A theoretical study has been made of various means to increase the smoothness of flight through rough air. This job order is to cover placing this analysis in a form suitable for a report, and preparation of a report on this study. In addition, measurements will be made in flight to verify the predicted response characteristics of airplanes, and bench tests will be made of servomechanisms intended for use with the automatic control device.

Justification: The preliminary analysis has indicated that the successful operation of a device to provide smooth flight in rough air requires careful selection of the design parameters, but that with a suitable design very promising results may be obtained. Such a device is not primarily intended to reduce stresses in the airframe due to severe abrupt gusts, but rather to reduce changes in acceleration which are primarily responsible for passenger discomfort and airsickness. The analysis indicates that current types of servomechanisms are capable of providing the desired rapid response of the controls. Inasmuch as a device of this type would be of great interest for airline operation, it is considered desirable to publish the results of the analysis and to continue this work with a view to eventual flight demonstration of such a device.

Note the manhour estimate. The value of 3000, about 1.5 man years, was just a guess based on the effort put into previous reports. The cost of $6000 was based on a standard rule of $2 per manhour. Little effort was required for this phase of the job approval.

When the work described in this job order was undertaken, I appointed Christopher C. (Chris) Kraft, Jr. as project engineer. Kraft was then an engineer in my section with experience in flying qualities and with work using the free-fall and wing-flow methods. Later in his career, he was a flight controller during the Apollo missions and was made director of the Johnson Space Flight Center following Dr. Gilruth's retirement. Kraft and I made additional calculations to have a logical series of examples to place in the report, which was later entitled *Theoretical Study of Some Methods for Increasing the Smoothness of Flight through Rough Air* (ref. 13.2). The report starts with a review of the available data on the causes of airsickness. The main source of data, which is still believed to be the best available, was a series of tests made during WW II at Wesleyan University, in which subjects were tested with various wave forms of vertical acceleration in an elevator. The results showed that relatively low-frequency variations of acceleration that had periods of 1.4 seconds and greater were the most important causes of motion sickness.

Excellent control of the directional and rolling disturbances of airplanes could be obtained with conventional autopilots available at the time the report was written. These devices, however, were relatively ineffective in reducing the vertical accelerations of airplanes. The report therefore concentrated on the problems of longitudinal gust alleviation and control.

Following a section on the theoretical analysis, examples were included in the report to show the use of elevator alone, flaps alone, and a combination of flap and elevator motion to offset the effect of gusts. Next, two types of gust-sensing devices, a vane ahead of the nose and an accelerometer in the airplane, were considered. Because these systems affect the controllability of the airplane, detailed studies were made of static and dynamic longitudinal stability of airplanes incorporating these systems.

In various studies of gust-alleviation systems made by other researchers since the one described herein, the now popular approach of optimal control theory has been applied. In this method, some balance is sought between the amount of control motion required by the system and the amount of reduction of acceleration achieved. In the study made by Phillips and Kraft, however, an effort was made to achieve complete gust alleviation within the limits imposed by the assumptions of the theory. Complete alleviation was found to be possible with the vane-type sensor, but not with the accelerometer sensor. Inasmuch as complete alleviation was found to be possible with reasonable control motions, the use of optimal control theory when a vane sensor is used is really not optimal and represents an inappropriate application of this theory.

The theory used for the analysis is very similar to that taught by Professor Koppen in my courses at MIT. This theory, in turn, was based on the theory first presented by G. H. Bryan in England in 1903 and later in the textbook *Stability in Aviation* in 1911 (ref. 4.1). This theory shows that the longitudinal and lateral motions of the airplane can be considered separately and assumes small disturbances so that all aerodynamic forces and moments can be assumed to vary linearly with the magnitude of the disturbance. This theory was extended to calculate the response of an airplane to gusts and reported in NACA Report No. 1 and later reports in the period 1915–1918 by Edwin B. Wilson, then a professor of Physics at MIT. To study the effect of a gust-alleviation system, it was necessary only to add to Wilson's theory the additional forces and moments caused by the airplane's control surfaces as they were moved by the gust-alleviation system. Like Wilson, I considered that the gust was constant across the span, though this subject was studied in more detail later.

Two refinements were added to the theory, as presented by Wilson, that were found to be of importance for the study of gust-alleviation systems and that are believed to make the results very close to what would be obtained with an exact computer simulation such as would be possible with modern electronic computers. First, the penetration effect was considered, that is, the difference in the time of penetrating the gust by a vane ahead of the nose, the wing, and the tail. In Wilson's study, the gust was assumed to affect all parts of the airplane simultaneously. Second, the time lead or lag effects caused by gust penetration were approximated by a linearized representation to keep the equations linear. This technique had been used by Cowley and Glauert in a British report published in 1921 to improve the calculation of pitch damping of airplanes by taking into account the time for the downwash leaving the wing to reach the tail (ref. 13.3). A similar method was used in the present analysis to account for all the lead and lag effects, such as the lead of the vane in penetrating the gust, the lag in response of the servomechanism operating the flaps, and the lag of downwash from the wing and flaps in hitting the tail, in addition to the lag of the gust itself in reaching the tail. These lead and lag effects were found to be very important, particularly in affecting the damping of the short-period longitudinal motion of the alleviated airplane.

A useful advantage was found in this approach in that all the effects of the alleviation systems studied could be considered as changes to stability derivatives of the basic airplane. Most of the effects of these derivatives were known from experience, or at least had simple physical interpretations. In addition, the order of the equations was not increased over that of the basic airplane.

These equations made it possible to solve for the characteristics of the system that would produce complete gust alleviation, that is, that would produce zero response to a gust. In examining these formulas, I suddenly realized that the results had a simple physical

interpretation. This interpretation is shown in figure 13.2, in which an airplane with a vane-type gust-alleviation system is shown penetrating a region in the atmosphere where there is a change in vertical gust velocity (called a step gust). First, the vane is deflected by the gust. If the servomechanism operating the flaps has a time lag equal to the time for the gust to reach the wing, then the flap moves just at the right time and the right amount to offset the lift change on the wing due to the gust. For the airplane response to be zero, however, the pitching moment about the center of gravity caused by the flap deflection must be zero. This condition is not ordinarily obtained with conventional wing flaps, but may be obtained by moving the elevators the correct amount in phase with the flaps. A little later, the tail is affected by the gust and by the downwash from the flaps. The effect of the forces on the tail can be eliminated in two ways. Either the flap downwash must be equal and opposite to the effect of the gust on the tail or the elevator must be given an additional movement to cancel the combined effect of the flap downwash and the gust.

The simple interpretation of the action of a perfect gust alleviation system has interesting ramifications. First, the influence of the feedback of angle of attack from the vane makes the alleviation system a closed-loop control system, but the feedback decreases as the system approaches the condition of complete gust alleviation. In this condition, the system behaves as an open-loop control, because there is no motion of the airplane to be sensed by the vane. Second, the consideration of lag effects is seen to be exact in the limiting case of perfect alleviation, even though these results were obtained from an approximate linearized theory. Third, the discovery of this interpretation could presumably have been made a priori, without use of any theory, but in my case, working through the theory first and examining the resulting formulas was necessary for me to realize that this interpretation existed. Many simple physical principles in the history of physics

FIGURE 13.2. Concept for complete alleviation of vertical gusts shown by the theoretical analysis. Note that the elevator moves in phase with the flaps, offsetting the flap pitching moment. Flap downwash, with opposite direction from normal, offsets the lift on the horizontal tail due to the gust.

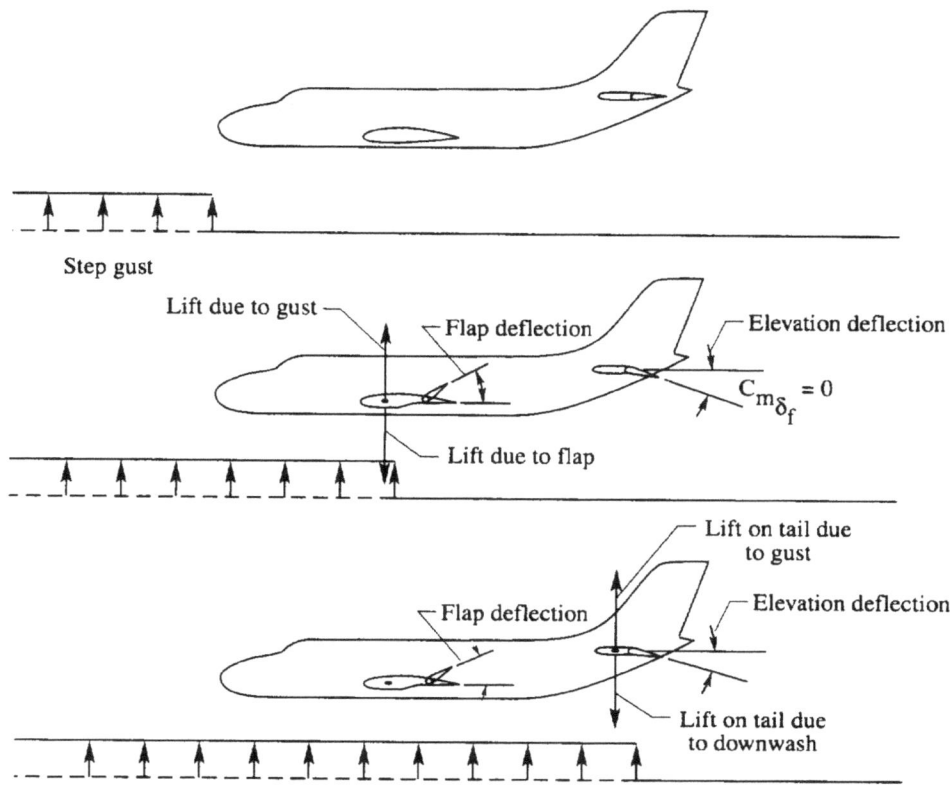

Step gust

Lift due to gust — Flap deflection — Elevation deflection
$C_{m_{\delta_f}} = 0$
Lift due to flap

Lift on tail due to gust
Flap deflection — Elevation deflection
Lift on tail due to downwash

have been realized only after long periods of thought and analysis by their discoverers.

The vane-type system studied has some disadvantages when adjusted to give complete alleviation. The system results in an airplane with zero lift and zero pitching moment due to angle of attack. The airplane would respond to pilot's commands as provided by the direct lift control system, but would have no inherent tendency to stabilize in a new equilibrium condition. For this reason, systems were studied that retained a small amount of longitudinal stability. Use of these systems was found to be feasible and did not seriously reduce the gust-alleviation properties of the perfect system.

The studies of the use of an accelerometer to sense the gusts showed that the gain of the system, that is, the amount of flap deflection used for a given change in acceleration, had to be limited otherwise a poorly damped

short-period vertical oscillation would result. As a result, the amount of alleviation potentially available with this system was limited. On the other hand, the use of the accelerometer sensor inside the airplane avoids the problem of having a delicate vane exposed to potential damage from handling. In practice, the system with an accelerometer sensor would be much more likely to excite structural oscillations of the airplane, thereby further limiting the gain and requiring a more detailed analysis to insure the safety of the system.

Design and Test of a Gust-Alleviated Airplane

During the publication process of the report on the analysis, work was started on a

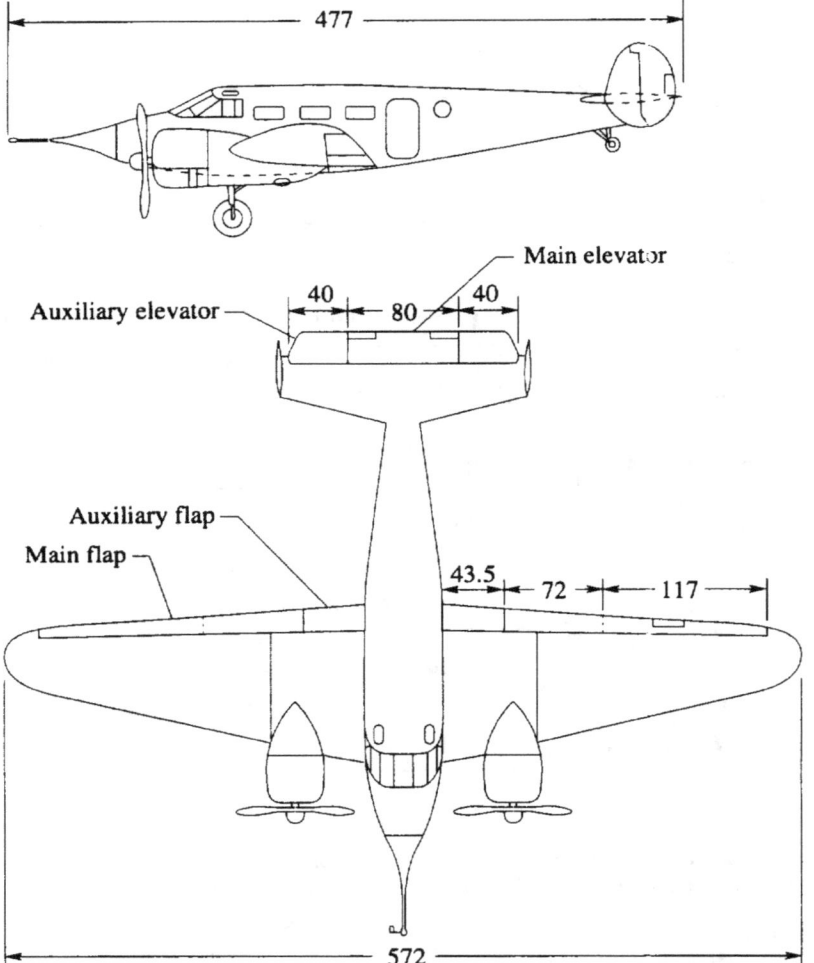

FIGURE 13.3. Two-view drawing of the Beech B-18 test airplane showing the modified control surfaces. Dimensions in inches.

program to demonstrate gust alleviation in flight. The airplane chosen for the program was a Beech B-18, a small twin-engine transport. The airplane was obtained from the Navy and had the Navy designation C-45. Because of the need for some major alterations to the airplane control surfaces, Kenneth Bush and Edwin C. Kilgore of the Engineering Division were called into the project to do the design work. Steve Rock of the Instrument Research Division was assigned to design the servomechanism to actuate the control surfaces.

At the time the design was started, about 1950, electronic control systems had a poor reputation for reliability. These systems used vacuum-tube amplifiers. Furthermore, the use of techniques of redundancy to improve reliability had not then been developed. For these reasons, many of the design features were governed by safety considerations. All autopilots in use at that time were designed so that the pilot could readily overpower the autopilot with his manual control system in the event of a failure. This method could not be used with the gust-alleviation system because the wing flaps, which required large

FIGURE 13.4. Nose boom and angle-of-attack vane installation on Beech B-18 test airplane.

operating forces, were not normally connected to the pilot's control stick. As a result, the design features described in the following paragraphs were incorporated.

A drawing of the airplane as modified for the gust-alleviation project is shown in figure 13.3. A boom was built on the nose to hold the angle-of-attack vane as shown in figure 13.4. The wing flaps, which normally deflect only downward, were modified to move up and down because both up and down gusts must be counteracted by the system. The elevator was split into three sections, the two outboard segments being linked to the flaps for use with the gust-alleviation system and the inboard segment being used in the normal manner for pitch control (figure 13.5). Finally, small segments of the flaps near the fuselage were driven separately from the rest of the flap system so that they could be geared to move either in the same direction or in the opposite direction from the rest of the flaps (figure 13.6).

As stated previously, perfect gust alleviation according to the theory could have been obtained either by driving the elevator separately from the flaps and with a different phase relationship or the elevator could have been geared directly to the flaps and the downwash from the flaps altered to offset the gust at the tail. The latter method was selected for the following reasons. With a direct mechanical linkage between the flaps and the outboard elevators, the pitching moment due to flap deflection, a critical quantity for longitudinal stability, could be finely adjusted as required and would hold its setting. If a separate servomechanism and electronic amplifier had been used to operate the elevators, the gain of the amplifier might have drifted and caused the airplane to become unstable. In fact, the gain of the amplifier between the vane and the flaps often did vary in flight by amounts that could have caused violent instability if a similar

FIGURE 13.5. Three-quarter rear view of Beech B-18 test airplane showing elevator control split into three segments.

FIGURE 13.6. Three-quarter rear view of Beech B-18 test airplane showing modified flap system with oppositely deflected inboard segment.

amplifier had been used to operate the elevators.

The small inboard segments of the flaps were used to reverse the direction of the flap downwash at the tail, as required by the theory if the elevators moved in phase with the flaps. This method of course, reduces the flap effectiveness in producing lift. To regain sufficient flap effectiveness, the flaps and ailerons were geared together so that the flaps and ailerons deflected symmetrically for gust alleviation. These surfaces were driven by an electrical input variable-displacement pump hydraulic servomechanism, which was taken from a naval gun turret. Electrical signals from the vane and the pilot's control column were combined in a vacuum-tube amplifier and fed to the control valve of the servomechanism. For lateral control, the entire flap and aileron system was deflected asymmetrically through a separate servomechanism of the same type. The system was designed so that the airplane could be flown through its original manual control system if the alleviation system failed or was switched off. In this configuration, the control wheel on the pilot's side remained connected at all times to the inboard segment of the elevator and to the ailerons. In the gust-alleviation mode, the ailerons were driven symmetrically through preloaded spring struts. This system remained connected in the manual mode, but the pilot could overpower the forces in the preloaded struts with his inputs to the control wheel. With the system in the gust-alleviation mode, the control wheel on the copilot's side was used to apply control inputs through the electronic control system. In addition, it remained connected through the mechanical linkage to the inboard portion of the elevator.

When the pilot turned the alleviation system off, the actuator driving the flaps was bypassed, and a separate hydraulic actuator with its own accumulator forcibly drove the flaps to neutral with a caliper-like linkage that could capture the flaps in any position.

Changes in angle of attack due to change in airspeed or drift of the amplifier used to operate the flaps could have caused the trim position of the flaps to vary slowly in flight. To maintain the trim position of the flaps at zero deflection over long periods, a mechanical ball-disk integrator driven by the flap linkage was used to feed an additional signal into the flap servomechanism to slowly run the flaps to the neutral position. This system had a time constant of 10 seconds, which was slow enough to avoid interference with the gust-alleviation function or the control of the airplane. To provide automatic control of the lateral and directional motion of the airplane in the gust-alleviation mode, a Sperry A-12 autopilot was connected to the aileron and rudder systems. This autopilot was the most advanced type available at the time of the tests.

Finally, the pilots landed the airplane with the wing flaps in neutral. This operation did not pose any problem on the long runways at Langley Field, but the design of a high-lift flap system that can also provide upward deflection of the flap for gust alleviation

remains one of the engineering problems of such systems that no one has yet tried to solve.

The airplane was instrumented with strain gauges to measure wing shear and bending moments at two stations and tail shear and bending moment at the root. The project became a joint project with the Aircraft Loads Branch. After the tests were completed, two reports had been published by each group: an initial and a final report on the gust-alleviation characteristics and an initial and a final report on the loads (refs. 13.4, 13.5, 13.6, and 13.7).

The tests occupied a long period of time. One of the main problems encountered was finding suitable rough air. The NACA test pilots were understandably conservative in flying experimental airplanes and usually declined to do test flying in clouds or stormy weather. Clear-air turbulence occurred on occasions after passage of a cold front. These conditions were used as often as possible in making the tests, but frequently the turbulence was of low intensity. For these reasons, some of the data were not as extensive as might have been desired.

Despite these problems, results were obtained with various sets of gearings to obtain varying degrees of longitudinal stability, and cases with the inboard flaps in neutral and moving oppositely from the outboard flaps were studied. A time history comparing the airplane motions in flight through rough air with the system on and off is shown in figure 13.7. In these runs, obtained early in the test program, the inboard flap segments were locked.

The results of later tests in which the inboard flap segments moved oppositely from the rest of the flap system are shown in figure 13.8. These results are shown as power spectral densities of the normal acceleration and pitching moment plotted on log-log scales. The results show that the system was effective in reducing the response in both normal acceleration and pitching velocity at frequencies below about 2 hertz. These plots

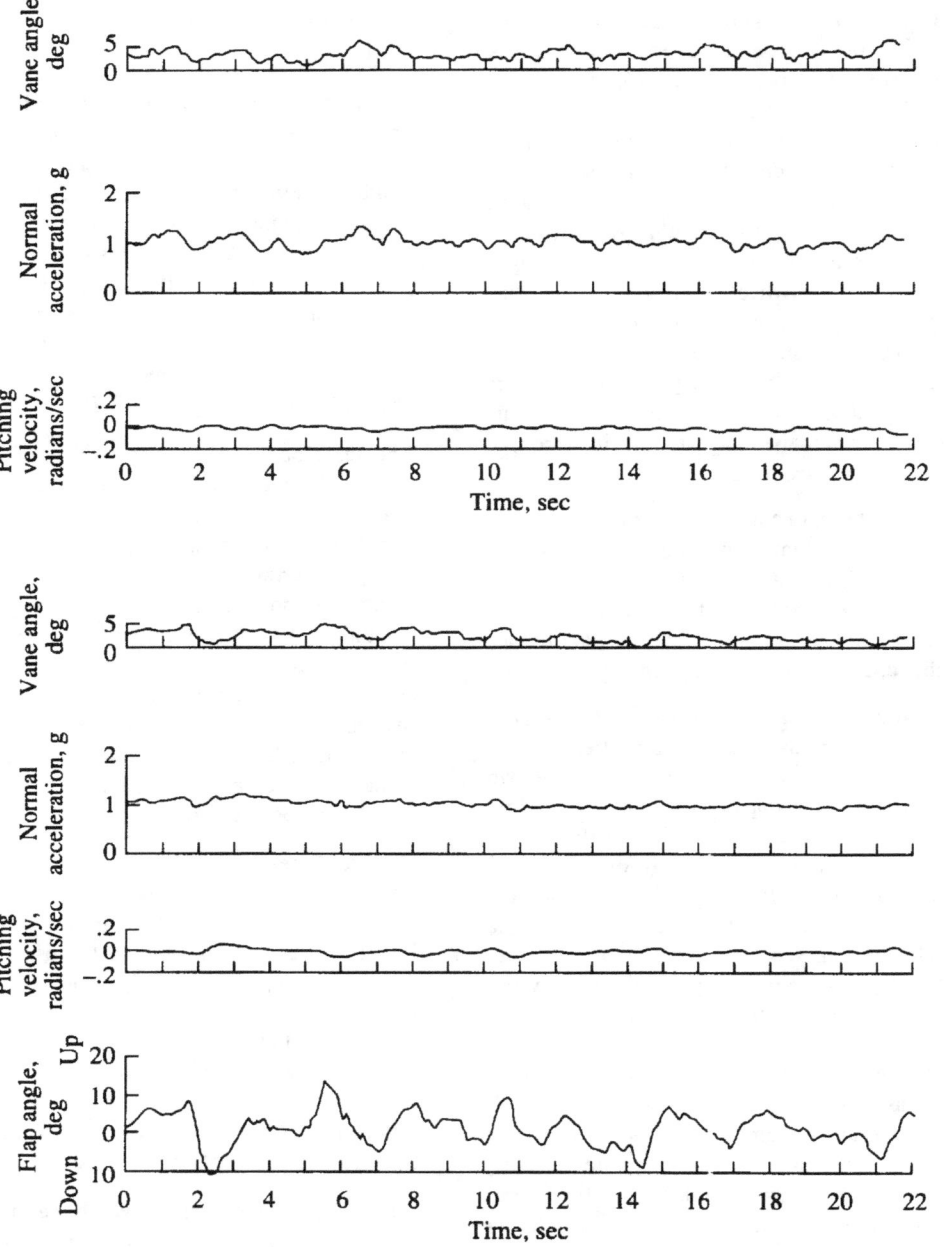

FIGURE 13.7. Comparison of flights in light turbulence, basic airplane, and gust-alleviated airplane. Initial configuration with center-section flaps in neutral.

(a) Basic airplane (top three).

(b) Gust-alleviated airplane (bottom four).

of power spectral density are the results of the usual evaluation procedure for randomly varying quantities, but they do not give a very clear comparison of the results obtained with the alleviation system on and off. The power spectra present the square of the recorded quantities, which tend to exagger-ate the differences, while plotting on log-log paper tends to reduce the apparent differ-ences. The question therefore arises as to how the data could be compared to give an impression of the effectiveness of the system more meaningful to the user. For this reason, data on the normal acceleration responses

FIGURE 13.8. Comparison
of power spectral
densities of normal
acceleration and pitching
velocity for basic
airplane and alleviated
airplane. Center-section
flaps deflected oppositely
from outboard flaps.

(a) Normal acceleration
(left).

(b) Pitching velocity
(right).

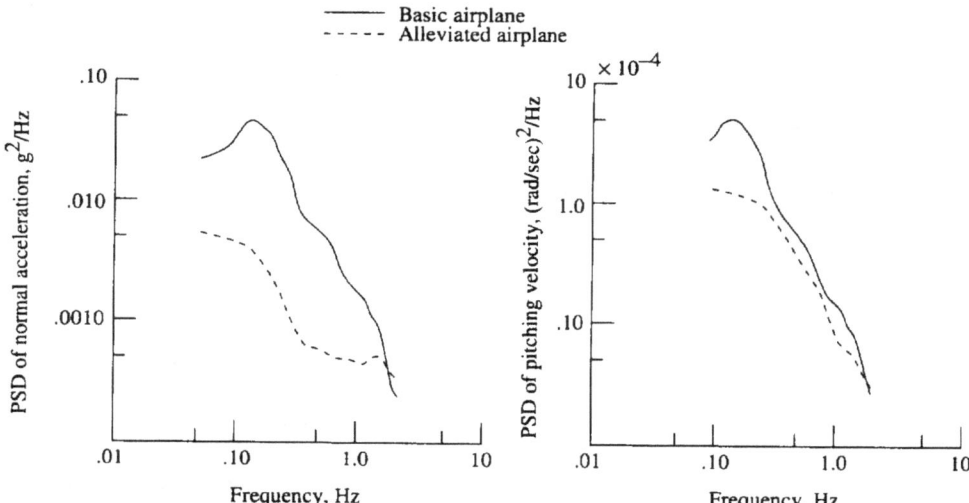

FIGURE 13.9. Comparison
of normal acceleration of
basic airplane and
alleviated airplane
shown on linear scales on
two different types of
plots.

(a) Amplitude of
transfer function of
(n/α_g), or ratio of
normal acceleration
to gust angle of
attack, as a function
of frequency (left).

(b) $\sqrt{PSD}$ of normal
acceleration, or
amplitude the
component of
normal acceleration
at each frequency, as
a function of
frequency (right).

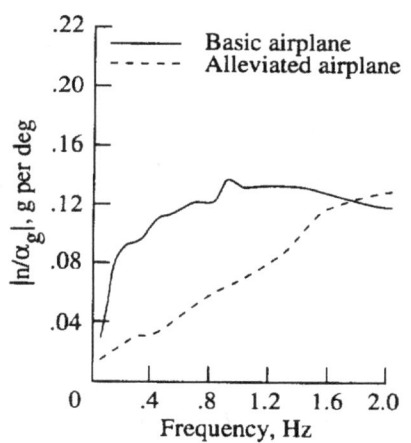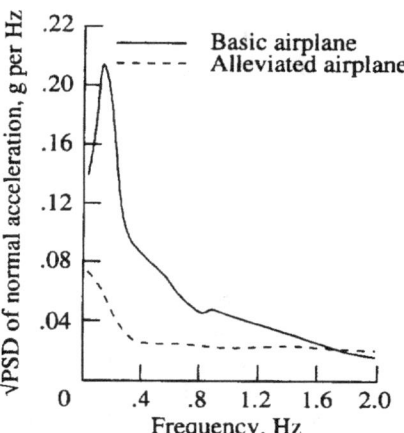

are plotted in two different ways in figure 13.9. In part (a) of this figure, results are plotted on linear scales in the form of a transfer function, that is, the ratio of normal acceleration to gust angle of attack for sinusoidal inputs of various frequencies. This plot is of interest to the control engineer and

shows correctly the relative magnitude of the normal acceleration for the two cases at various frequencies, but does not include the variation of the actual gust forcing function with frequency. In part (b), the square root of the power spectral density of the response is plotted as a function of frequency on linear scales. This plot attempts to show the actual magnitude of the normal acceleration at each frequency. This plot is believed to be more meaningful in interpreting the passengers' impression of a ride in the airplane.

As can be seen, the basic airplane has a large peak in the acceleration response at a frequency about 0.2 hertz. This low-frequency peak is typical of the response of unalleviated airplanes and occurs because of the larger amplitude of the turbulence input at low frequencies. In the case of the alleviated airplane, the response is reduced to a fairly constant, relatively low value in the frequency range between 0 and 1.6 hertz. For the degree of turbulence encountered, this reduction greatly improved the subjective impression of ride comfort. Though the results are not shown on the plots, the response at frequencies above 2 hertz with the alleviation system operating were slightly increased above those of the basic airplane. Also, in turbulence of relatively large magnitude, the pilots noted a fore-and-aft oscillation caused by drag of the flaps at large deflections.

Though the results obtained were quite gratifying, the question arises as to why the performance of the system was not better, inasmuch as the theory predicted perfect gust alleviation. One of the main reasons was the nonlinear characteristics of the servomechanism used to drive the flap system. The designer of the system in the Instrument Research Division had been requested to provide a rather sharp cutoff in the response beyond 2 hertz, to avoid the possibility of exciting wing flutter. The C-45 was a very stiff airplane, with the wing primary bending mode at 8 hertz. It was intended, therefore, that the output of the servomechanism should be close to zero at a frequency of 8 hertz. It was not until after the device was installed that it was found that the cutoff in response had been obtained by rate limiting the output. This provided a steep cutoff, but the response was a function of amplitude. The result was that in a large amplitude gust, rate limiting was encountered and the allevi-

ation was reduced just when it was needed most. A later unpublished investigation of the effects of rate limiting demonstrated these adverse effects and showed that with sufficiently severe rate limiting, a continuous sawtooth oscillation of the flap would be encountered in rough air, which resulted in response greater than that of the basic airplane.

A second problem was the nonlinear lift characteristics of the flaps. The flaps deflected about plus or minus 25 degrees, but the slope of the curve of lift versus flap deflection fell off markedly before this deflection was reached. As a result, a fixed gain between the vane and the flaps did not give a uniform value of the ratio of lift to gust angle of attack. This nonlinear response would be expected to introduce higher frequency harmonics into the response and probably accounts for the increase in response beyond 2 hertz. This and other characteristics were not known before the tests were made. I concluded that on any future project of this type, wind-tunnel tests of the airplane to determine the control characteristics would be advisable. Despite these deficiencies, enough was learned to show the feasibility of gust alleviation and to show how the results could be improved in a future attempt.

Analytical Studies of Additional Problems

The studies of gust alleviation introduced several problems that had not been considered previously in airplane design. One problem was the effect of variations in gust velocity across the wing span. The use of a single vane on the center line of the airplane to sense the gusts would work perfectly if the gust velocity were constant across the wing span, but would be less effective if different values of gust velocity were encountered at different stations along the span. This problem can be studied if the turbulence in the atmosphere is assumed to be isotropic (or for this application, axisymmetric); that is, it has the same characteristics regardless of the direction in which the airplane is flying. This assumption appears reasonable for most types of turbulence. The theory of isotropic turbulence had been studied theoretically. The nature of the spectrum of turbulence, that is, the way the gust velocity varies with gust wavelength, had been determined. The gust velocity has been found experimentally to vary approximately directly with the wavelength even for wavelengths many times larger than the wing span of the largest existing airplanes. The most intense gusts, which cause the most disturbance to the airplane, have wavelengths long compared to the wing span and are therefore approximately constant across the span. Gusts with wavelengths short compared to the span have low intensity and therefore do not disturb the airplane much. The use of a single vane on the center line is therefore quite effective. Gusts of wavelength short compared to the span may cause up and down loads that average out across the span. For these gusts, the response of the vane at the center line would be too large. These gusts have such high frequency, however, that the vane response is filtered by the lag in response of the flap servomechanism. For a vane located ahead of the nose, the lag in the flap servomechansm is also beneficial in delaying the response sensed by the vane until the gust reaches the wing. Considering these factors, the effectiveness of a gust-alleviation system using a vane on the centerline may be shown to be about 98 percent as effective in axisymmetric turbulence as it would be with gusts constant across the span. This problem was studied in more detail after the flight investigation was completed (refs. 13.8 and 13.9).

An interesting optimization problem, which to my knowledge has not been solved, is to determine the optimal filter to place between the vane and the flap to obtain the greatest amount of gust alleviation, considering the distance of the vane ahead of the wing, the wing span, and the spectrum of atmospheric turbulence. This problem is only of academic importance, however, as

about 98 percent alleviation was obtained by using a second-order linear filter with reasonable frequency and 0.7 critical damping (refs. 13.6 and 13.7).

Some time after completion of the tests, a summary report was given as part of a lecture series at Renssellaer Polytechnic Institute (ref. 13.10). This paper gives a more complete discussion of the subject of gust alleviation than contained herein.

Future Possibilities of Gust Alleviation

Some review of later efforts in the field of gust alleviation may be of interest, inasmuch as the development of computers and automatic control technology would permit approaches quite different from that used in the early NACA tests. It was a disappointment to me that very little effort was made by aircraft companies to incorporate provision for gust alleviation even after the development of control technology would have made it more feasible.

To my knowledge, the only airplane in service that incorporates a system performing some gust-alleviation function is the Lockheed 1011. In some later models, the wing span was increased by extending the wing tips to allow the airplane to carry greater loads. To avoid changing the wing structure to withstand greater bending moments, the ailerons were operated symmetrically by an automatic control system to reduce bending moments due to gusts.

One reason for the lack of interest in gust alleviation is that following the NACA tests, jet transports were introduced. As a result of higher wing loading, swept wings, flight at higher altitudes, and the use of weather radar to avoid storms, these airplanes were much less likely to encounter violent airplane motions that would cause airsickness. In addition, the problem of gust alleviation became more difficult because the structural flexibility of these airplanes placed their structural frequencies closer to the frequency range of interest for gust-alleviation. As a result, structural response would have to be considered in designing the system. In recent years, these reasons for avoiding the use of gust-alleviation systems have become less significant. Extensive use is now made of commuter airplanes that fly at lower altitudes and frequently encounter rough air. In addition, methods have been developed to analyze the structural response and to damp out the structural modes by use of automatic control systems.

Review of Work by René Hirsch

In closing, a brief review is given of the work of René Hirsch, whose thesis was mentioned at the beginning of this chapter. Also, a few programs and studies applicable to gust alleviation that have occurred since the NACA program on the C-45 are reviewed.

After reviewing Hirsch's thesis, no more was heard of his activity until the early 1950's, during the course of the NACA program. At this time, a French report was discovered revealing that, following WW II, Hirsch had made additional wind-tunnel tests in the French large-scale tunnel at Chalais-Meudon on a model of a proposed airplane and had built this small twin-engine airplane incorporating his system. The airplane was envisioned as a quarter-scale model of a piston-engine transport of a class similar to the Douglas DC-6 or the Lockheed Constellation, which were the largest transports in service in that period. Correspondence was established with Hirsch and additional reports and information were obtained (ref. 13.11). Hirsch's airplane had a wing span of 27 feet and had two 100-horsepower motors. It incorporated the same system described in his thesis, in which the halves of the horizontal tail moved on chordwise hinges to operate flaps on the wings. The conventional elevators provided not only pitching moments, but moved the tail halves

about their chordwise hinges to cause the flaps to move in the direction to provide direct lift control. In this way, the loss of longitudinal control due to the gust-alleviation system was overcome. Hirsch's airplane incorporated many other ingenious features, including provisions for reducing rolling moments due to rolling gusts and lift due to horizontal gusts. His design also incorporated large pneumatic servos operated by dynamic pressure to restore damping in roll and to stabilize the rate of climb or descent. The airplane had good handling qualities and appeared to have been very successful in providing a smooth ride, as shown by some time histories in rough air with the system turned on and off. After about 30 flights, the airplane ran into a ditch at the end of the runway and was damaged.

No more was heard until 1967, when another report appeared showing that the airplane had been rebuilt and equipped with two 180-horsepower motors (ref. 13.12). In this condition, the airplane made numerous additional flights with somewhat more complete instrumentation. A photograph of the airplane in flight is shown in figure 13.10. From the data obtained, the results appeared very similar to those obtained with the NACA C-45 in that the accelerations due to gusts were reduced by about 60 percent at frequencies below about 2 hertz, but were increased somewhat at higher frequencies.

All of Hirsch's work in designing and building his airplanes was done with his own funds, though some help with instrumentation was obtained from ONERA, the French equivalent of the NACA. Outside of France, Hirsch's work was little known. His reports described the work in general, but were not sufficiently detailed to give engineering data on all of the ingenious ideas and systems incorporated on his airplanes.

In 1975 during a trip to France, I visited Hirsch. He had found at that time that a twin-engine airplane was too expensive for him to operate during the oil crisis, and he had donated it to the French Air Museum. When

FIGURE 13.10. Flight photograph of René Hirsch's first airplane as equipped with larger engines.

FIGURE 13.11. René Hirsch in front of his Aerospatial Rallye light airplane equipped with gust-alleviation system.

I saw the airplane, it stood like a little jewel amid a group of dilapidated antique airplanes in an old WW I hangar at Villaroche. It is now on display at the new French Air Museum at Le Bourguet.

At that time, Hirsch was starting modification of a single-engine light plane, the Aerospatial Rallye, to incorporate his gust-alleviation system. This airplane was completed and flying when I visited France again in 1980. Hirsch is shown standing in front of his Rallye airplane in figure 13.11. This airplane was never as successful, in Hirsch's opinion, as the first one, probably because of its lower airspeed and lower frequency of response of the flap systems. In recent years (1995), Hirsch modified a third airplane, the Sobata Trinidad, with small canard surfaces ahead of the wing root to operate the flaps. The more forward position of these sensing surfaces was intended to improve the response to high-frequency gusts. Hirsch died in August 1995 at the age of 87 without having had the opportunity to test his latest design.

Hirsch's dedication to the pursuit of gust alleviation is a remarkable story in view of the general disregard of this subject by the rest of the aviation industry. Of course, the aeromechanical systems used by Hirsch have been superseded by automatic controls using computers and electrohydraulic actuators. Hirsch readily admitted that he would prefer such systems but was unable to afford them.

Later Studies by Other Investigators

Though a number of studies of gust alleviation have been made during the years since the C-45 tests, most of them have not contributed any notable new developments. Only two are mentioned to bring the subject up to date. One is the so-called LAMS project, an acronym for Load Alleviation and Mode Stabilization, conducted at the Air Force Flight Dynamics Laboratory at Wright-Patterson

Air Force Base, Ohio, about 1968–1969 (ref. 13.13). In this project, a B-52 bomber was equipped with an electronic analog-type flight control system to operate the existing flight controls to damp out structural modes. This work is important because consideration of damping of structural modes would be required in any attempt to install a gust-alleviation system in a high-speed airplane. The report illustrates the success of modern control analysis techniques (as they existed at that time) in designing a modal damping system and in predicting the results obtained.

The second contribution of note is the analytical work of Dr. Edmund G. Rynaski, formerly of Calspan and now at EGR Associates, in designing a system to alleviate both the rigid-body motions and selected structural modes of an airplane (ref. 13.14). Rynaski's work, based on matrix analysis techniques, shows how to provide essentially open-loop control of the rigid-body modes, as was done on the C-45 airplane, as well as to provide open-loop cancellation of a selected number of structural modes. This method also makes possible improved damping of higher order structural modes by use of closed-loop control.

With the availability of digital flight computers and modern control actuators, different approaches should be considered for gust alleviation. One approach would be to operate the flaps on the wing as a function of angle of attack sensed by a vane or similar device, but to operate the elevators by a modern reliable pitch damper as part of a longitudinal command control system to control the pitching response of the airplane. Another approach would be to calculate absolute gust velocity on line by a method similar to that referred to previously in the work by Crane and Chilton in measuring gust velocity (ref. 12.5). This method requires correcting the angle of attack measured by a vane for the inertial motions of the airplane at the vane location. This signal could be used as an input into a gust-alleviation system without the need to modify the normal control or handling qualities of the airplane.

CHAPTER 14

Lateral Response to Random Turbulence

The field of flight dynamics has been studied since the early days of aviation. As mentioned previously, G. H. Bryan in England worked out the complete equations of airplane motion the same year as the Wright brothers' first flight (ref. 4.1). By the 1950's, it was difficult to find any field of research in flight dynamics that had not been thoroughly explored. The gust-alleviation program, however, revealed one problem that had received very little attention. This problem was the lateral response (that is, rolling and yawing response) due to flight through atmospheric turbulence.

Modeling the Turbulence

As mentioned in the section on gust alleviation, the scale of turbulence in the atmosphere is ordinarily so large that the gusts may be considered constant across the wing span. This circumstance made it possible to use a single vane in the nose of the airplane to sense the gusts with the possibility (in theory at least) of getting 98 percent complete alleviation. In considering lateral response, however, such an assumption is obviously incorrect, because if the gusts were constant across the span, the airplane would receive no rolling disturbance due to turbulence.

The method of calculating lateral response of an airplane to discrete gusts was the subject of a report by R. T. Jones (ref. 5.3). He showed that the conventional equations of motion could be used to calculate the response to a rolling gust by assuming that a rotational motion of the atmosphere about the longitudinal axis acted on the airplane through the so-called stability derivatives associated with rolling of the airplane, that is, the rolling moment due to rolling velocity and the yawing moment due to rolling velocity. These derivatives were already used in the standard lateral stability equations and were well known as a result of previous theoretical and experimental research. The effect of a rotational motion of the atmosphere about a vertical axis could likewise be calculated with the aid of the yawing moment derivatives. What was needed was a way to determine the rolling and yawing motions of the atmosphere associated with random turbulence.

The turbulence in the atmosphere was ordinarily characterized by what is known as the point spectrum of turbulence. This spectrum is what would be obtained by measuring the velocity fluctuations along a straight line as it is traversed at a relatively high speed. Measurements of the point spectrum of vertical velocity, for example, were made by recording the motion of a vane to measured angle of attack and correcting the readings continuously for the inertial motion of the airplane at the location of the vane. This method was used in the measurements by H. L. Crane and R. G. Chilton described previously (ref. 12.5).

To extend this information to obtain velocity fluctuations across the wing span, the usual assumption made is that the turbulence is isotropic; that is, the spectrum of turbulence is the same no matter what direction the atmosphere is traversed. It seems reasonable, for example, that an airplane flying through a region of turbulence from north to south would experience the same amount of disturbance as an airplane flying through the same region from east to west. Using the assumption of isotropy, H. S. Ribner presented an analysis of response to turbulence using a two-dimensional Fourier series to represent the disturbances (ref. 14.1). This approach was also described in a textbook by Bernard Etkin (ref. 14.2). Franklin W. Diederich, in a doctoral thesis at California Institute of Technology under R. W. Liepmann, calculated the spectrum of lift on a wing flying through turbulence by using a correlation function approach. This thesis was later published as an NACA Technical Report (ref. 14.3). Diederich also showed how to calculate rolling and yawing moments on the wing, but the complete calculations were not carried out. I considered the approach used by Diederich to be easier to comprehend than that used by Ribner, but both should give the same answer if used correctly.

Calculation of Response

I assigned an engineer in my division, John M. Eggleston, to work with Diederich and complete the calculations of yawing and rolling moments (ref. 14.4). These calculations were made for wings of rectangular, elliptic, parabolic, and triangular planform and involved rather complex mathematical manipulations. Eggleston continued this work to make a complete matrix formulation

of the airplane response and added the effects of the fuselage and tail to those of the wing by a method that I suggested (ref. 14.5). Eggleston calculated the effects of the rolling moments, yawing moments, and side forces separately and then added these contributions to get the response spectrum of the resulting motion. An important consideration in these calculations is that the lateral motion is only statistically, rather than uniquely, determined in flight through any given velocity pattern along the fuselage centerline because different velocity distributions across the span may be encountered for a given distribution along the fuselage centerline. The calculations can therefore produce only the power spectrum of the response for a given point power spectrum of turbulence.

I continued my analysis to determine the relation between the spectra of effective rolling gusts and yawing gusts and the point spectrum of turbulence. Again, the results of Eggleston and Diederich's calculations (ref. 14.4) were used, but the effective atmospheric rolling and yawing motions were calculated and applied directly as inputs to the lateral equations of motion. I worked out some of the same examples used by Eggleston. Much to my surprise, the results turned out to be identical to those of Eggleston. These results were obtained with much shorter calculations, and to me, gave a clearer picture of the physical process involved (ref. 14.6). Later, a technical report was published presenting both Eggleston's matrix calculations and my shorter method (ref. 14.5) together with a simple 45-step procedure for calculating the power spectra of the lateral responses for specific cases.

The relation between the spectrum of rolling gusts and the point spectrum of turbulence used in the analysis is shown in figure 14.1. Ordinarily, a power spectrum is presented as the square of the dependent variable plotted against the independent variable on log-log paper. This type of plot is shown in part (a) of the figure. A squared variable is compatible with the statistical significance of the

FIGURE 14.1. Relations between the power spectrum of rolling gusts and the point spectrum of turbulence for various values of the ratio of scale of turbulence to wing span.

(a) Ratio of power spectrum of rolling gusts to point spectrum of turbulence versus reduced frequency as conventionally plotted on log-log scales (top).

(b) Square root of $(D\phi_g/\beta_g)^2$ as a function of frequency plotted on linear scales (bottom).

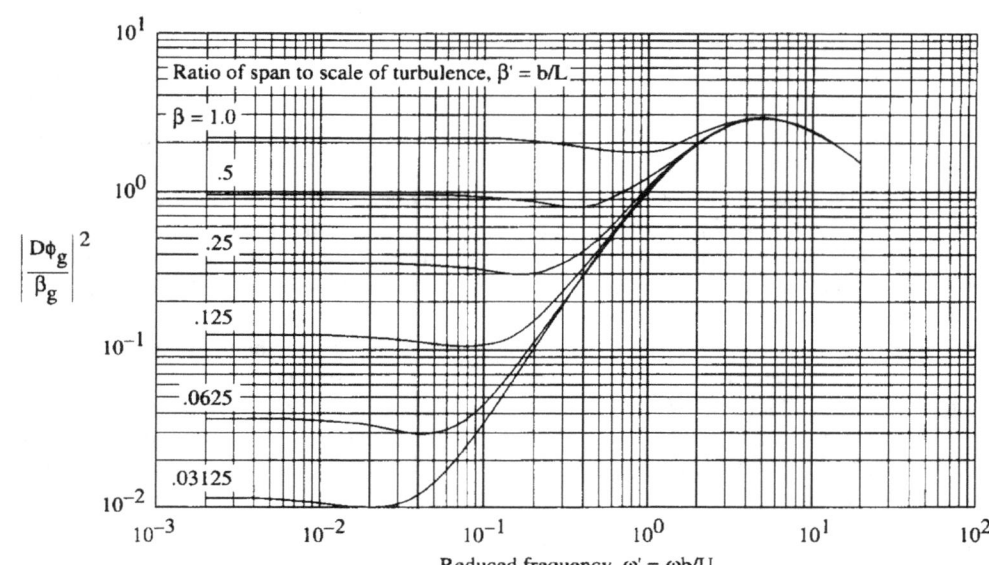

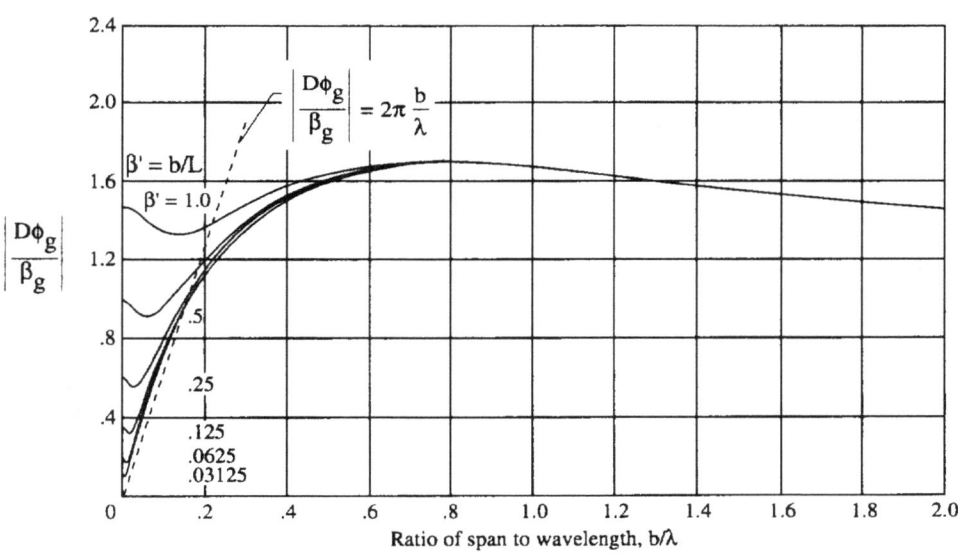

quantity in which only the amplitude and not the phase relationship is obtained from the analysis. Log-log paper is useful in showing a large range of the variables. This method of presentation, however, obscures a clear impression of the actual shapes of the curves. The same quantities shown by taking the square root of the ordinate and plotting it as a function of reduced frequency on linear scales is shown in part (b) of the figure. This plot shows the expected results that at very long wavelengths, the amplitude of the effective rolling gust is small because the gradient of vertical velocity along the span is small. At high frequencies, where the wavelength is small compared to the span, the effective rolling gust is again decreased because the fluctuations are averaged along the span. Between, the effects reach a maximum. In a region indicated by the sloping dotted line, the disturbances approach an approximation obtained by assuming a uniform gradient across the span equal to the slope of the central region of a sinusoidal gust with a wavelength corresponding to the frequency. The only rather puzzling aspect of the curve is the slight tendency of the curves to turn up at very low frequencies. Possibly this effect is caused by rare encounters with widely spaced large-amplitude rolling gusts, first in one direction and later in the other, which would introduce a harmonic of very low frequency in the spectrum.

At the time these studies were made, about 1958, I wanted to conduct some flight tests to verify the predicted relationship between the measured point spectrum of turbulence and the resulting lateral response of the airplane. These methods might have also provided an alternate method to measure the spectrum of atmospheric turbulence. Unfortunately, flight testing of high-speed airplanes at Langley was discontinued at that time. To my knowledge, such flight tests have never been carried out, and they remain one of the few uncompleted phases of flight dynamics.

The work of Franklin W. Diederich at Langley deserves special mention. His work, which involved both aeroelasticity and response to turbulence, came at a time when high-speed computers were just becoming practical and when closed-form solutions of complex problems were becoming less necessary. As a result, his work to produce widely useful charts and formulas appears to be largely forgotten, even by engineers working in the fields to which he made such valuable contributions. Now, when similar problems are encountered, a computer simulation of the specific problem under consideration is made with such techniques as finite-element methods or step-by-step solutions. The programs for making such calculations are generally available to the engineers involved. Though some familiarity with the physics of the problems is advisable, a detailed knowledge of classical theorems from aerodynamic theory or of methods of manipulating complicated mathematical functions is usually not required.

The work on response to random turbulence, including the subject of gust alleviation, has also received little attention from the aeronautical industry because the main structural design conditions on airplanes come from rare encounters with discrete gusts in thunderstorms. Such events have been studied through operational experience rather than by research programs designed to study the details of response to turbulence. As a result, much very interesting research work on this subject now lies buried in old technical reports that are occasionally unearthed by scholars, but rarely used in practice.

CHAPTER 15

Accident Investigations

The NACA and NASA were not routinely called in to give advice on all accident investigations, but in some cases in which the cause of a crash could not be readily determined or when a new experimental design was involved, the appropriate government organizations requested assistance. Accident investigations are important not only to correct defects in existing designs that might cause further accidents, but also to provide information for safer designs in the future. I was directly involved in a number of these investigations, some of which are sufficiently interesting to include in this autobiography.

In this discussion, the term "dynamic pressure" is used. The dynamic pressure is approximately the increase of pressure above the surrounding atmospheric pressure caused by the flow impinging on a flat surface normal to the airstream. It is given by the formula one half times density times the square of true airspeed. Such a value of pressure occurs at the nose of an open-ended tube facing into the airstream, called a pitot tube, and is used to measure airspeed as well as to determine loads on the airplane caused by the airflow. The term "indicated airspeed" is the airspeed shown by a standard airspeed meter when it is supplied with this pressure and a reference pressure called the static pressure, which is the true ambient pressure in undisturbed air. The instrument is calibrated to read true airspeed for the density at standard sea level conditions. The

true airspeed, required for navigation purposes, is greater than the indicated airspeed at altitudes above sea level because the density decreases with increasing altitude.

The abbreviation MAC stands for mean aerodynamic chord, a reference line based on the wing plan form that is used in deriving center of gravity location and in computing stability derivatives.

BAC-111 Crash Near Omaha, Nebraska, on August 6, 1966

From time to time, in investigations of crashes of civil aircraft, the NACA or NASA were contacted by the Civil Aeronautics Board (CAB), later called the National Transportation Safety Board (NTSB), to assist with accident investigations. One of these accidents that involved considerable study on my part was the crash of a BAC-111. This airplane was a small twin-engine jet transport, constructed by the British Aircraft Corporation (BAC), that was used rather extensively on short-haul routes in the United States (figure 15.1). In the course of the study, I attended a meeting with the British officials in England and other meetings in various locations in the United States, including Omaha, Nebraska, and Cleveland, Ohio.

FIGURE 15.1. The BAC-111 Transport Airplane.

The crash occurred when the airplane, which was flying at 5000 feet altitude (4000 feet above the terrain) and carrying 38 passengers and 4 crew members on a scheduled Braniff Airways, Inc. flight, burst into flames in the air and crashed in a field in a flat, upright attitude with no forward motion. As is usual in such crashes, there was little information available to determine the cause of the crash.

One of my first duties was to attempt to determine, from the locations of various pieces of wreckage around the crash site, the approximate altitude and airspeed of the sections at the time of breakup in the air. This kind of analysis requires calculating the trajectories of the various pieces as affected by their weight, drag, and surface area. In this case, the horizontal and vertical tail unit (a T-tail arrangement) and an outer wing panel were separated from the main wreckage. At that time, digital computers were available, but I had little experience in using them. I used a program called the Mimic program, which allowed the analysis to be conducted with the aid of a block diagram similar to that which would be used in making an analog computer solution and which had an unusually clear manual to describe its use. The solutions came out in good agreement with the expected altitude and speed of the airplane, which indicated that the airplane

broke up in normal flight without warning. Later, the CAB representative requested copies of the computer program to use in their analyses of other crashes.

All possible causes of the crash were investigated, both by BAC and by the CAB. These causes included such things as failure of the control system or feel devices, use of improper construction materials, use of incorrect design procedures or requirements, and fire due to fuel or hydraulic leaks. A contributing factor involved in many studies was that the airplane was flying through or above a roll cloud when it burst into flames and was a few miles away from an approaching line of heavy thunderstorms.

The cockpit voice recorder was recovered, but the only information on it was various noises, such as a rushing sound of air following the breakup and a sounding of various warning horns during the descent. There was no crash recorder such as is now required on commercial airplanes to record airplane motions or control positions.

In listening to the record on the voice recorder, it was noted that following the breakup, there was noticeable variation in pitch or "wow" as it is called by audio enthusiasts. I conceived the idea that the records obtained might be used to get some idea of

FIGURE 15.2. Rolling acceleration and rolling velocity of BAC-111 airplane as determined from voice recorder analysis.

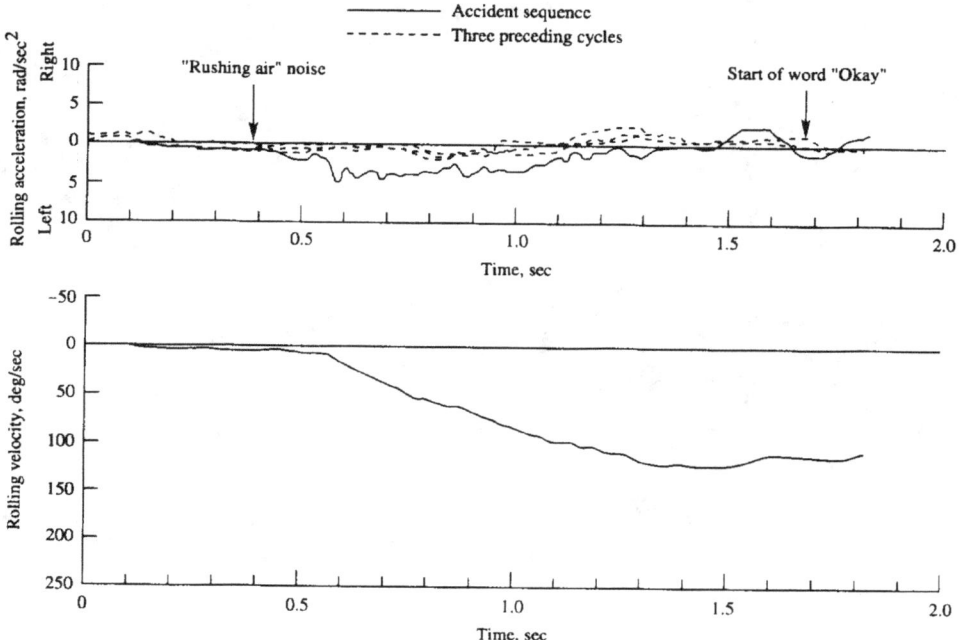

the motions of the airplane. The recorder had a flywheel in the tape drive that had its axis aligned with the longitudinal axis of the airplane. As a result, the speed of the flywheel would be affected by rolling accelerations of the airplane. A similar voice recorder was obtained for calibration tests in the NASA Instrument Research Division. The record obtained from the accident had a faint 800 hertz tone, probably a harmonic of the aircraft power supply, that formed an excellent time base. I laboriously analyzed the data, one cycle at a time, to determine the movement of the tape and was able to plot a record of rolling acceleration as a function of time. This record was then integrated to determine rolling velocity. The data showed an abnormally rapid rolling velocity to the left building up to over 100 degrees per second in less than 1 second following the start of the "rushing air" noise (figure 15.2). Though this record was interesting, it could not be relied on entirely because the tape speed was also affected by large values of linear acceleration.

To simulate the maneuvers following the breakup, engineers at the Langley 20-Foot Vertical Spin Tunnel had a crude model of the airplane built with the tail assembly and part of the left wing missing, as found on the wreckage (figure 15.3). This model was launched by a catapult from the top of the Langley Lunar Landing Facility (now Langley Impact Dynamics Research Facility), a girder structure 250-feet high, which for the scale of the model used, corresponded closely to an altitude of 4000 feet. The model would go into violent tumbling motion, but before hitting the ground would settle into a flat descent, sometimes without rotating and sometimes in a slow flat spin. The model always landed right side up, probably because of the wing dihedral.

Another interesting phase of the investigation was a number of lectures by a well-known meteorologist, Dr. Tetsuya Fujita of the University of Chicago, who discussed mesoscale turbulence and who had coined the term "downburst" to describe the downward rush of air cooled by heavy rain in a

FIGURE 15.3. **Model of damaged BAC-111 used for catapult tests to determine motion following breakup.**

thunderstorm. He pointed out that this air would spread out on the ground and would advance at a high speed several miles ahead of the main storm, which would result in roll clouds or in heavy clear-air turbulence.

The conclusion of the accident board was that there was no fault in the design or construction of the airplane and that the breakup had been caused by an abrupt gust with velocity well beyond the design requirements. Members of the BAC calculated that a gust of 144 feet per second angled at 45 degrees upward and 45 degrees from the right would have been required to break off the T-tail in the manner that occurred in the accident, whereas the design gust velocity at that time was 88 feet per second. Following loss of the tail, the airplane would immediately pitch down to large negative values of acceleration, which would cause the outer wing panel to break off.

The official accident investigation concluded that the airplane met all design requirements and that no changes in design were required. Warnings were issued concerning the possibility of high turbulence at low altitudes several miles from heavy thunderstorms. Subsequent experience showed that the con-

clusions were probably correct because no similar crash of a BAC-111 ever occurred. A complete copy of the National Safety Board report was published in *Aviation Week* (ref. 15.1).

I proposed a wind-tunnel study of turbulence of the type encountered in the accident, by using a large downward jet of air directed at the floor of a wind tunnel operating at very low speed. This study was never made. I have always wondered whether the airplane would have broken up flying anywhere into the storm front at the altitude and vicinity of the crash, or whether the gust was a single, isolated phenomenon with very small probability of being encountered by an airplane.

Crash of Convair B-58 Bomber

The Air Force started a program about 1952 to develop a supersonic bomber. The result was the Convair B-58 Hustler, a delta-wing aircraft with four engines mounted in nacelles under the wing (figure 15.4). The airplane was first flown in 1956. A large pod, almost as large as the fuselage itself, could

FIGURE 15.4. B-58
Hustler bomber airplane.

be carried under the fuselage. This pod was intended to be developed into a large nuclear weapon or a carrier for smaller munitions. The B-58 was a remarkable airplane in many ways in that it had much larger range, installed power, and load-carrying ability than any supersonic airplane developed before that time. By 1960, about 13 airplanes had been built. It is not surprising that, with such an advanced design, problems would be encountered in its development. By 1960, however, five airplanes had been destroyed in accidents, and the Air Force was engaged in a frantic effort to increase the safety of the vehicle and its operation.

Of the accidents that occurred, the first four involved such unrelated causes as a ground fire during fuel transfer, pilot error, a crash due to tire failure, and structural defects. The fifth accident, however, occurred in flight while an experienced test pilot was testing the engine control system. The airplane went to a large angle of sideslip and disintegrated in the air. This event might indicate a serious defect in the design. The Air Force set up their own investigating board and also set up a joint Air Force–industry board, of which I was a member.

Several of the board members were men I had worked with previously on the Society of Automotive Engineers (SAE) A-18 com-

mittee, known as the SAE Aerospace Control and Guidance Systems Committee. This committee was a remarkably informative organization of the top control engineers of various companies, equipment manufacturers, and universities. Meetings were held four times a year, usually at very attractive locations. The members had very few inhibitions on discussing their companies' latest projects because it was realized that a free interchange of ideas would stimulate development. Another feature that encouraged attendance was that no formal proceedings were published. Instead, each speaker brought enough copies of his presentation to a meeting so that each member could have one. As a result, the attendees immediately had a stack of useful information on the latest developments.

Two of the men prominent in the SAE committee who assumed leadership roles in the accident investigation were Duane McRuer and Irving Ashkenas, officials of a small company called STI (Systems Technology Incorporated) whose business was studying stability problems of airplanes. I was assigned the job of investigating the longitudinal stability, and Irving Ashkenas headed the study of lateral and directional stability. The Convair Corporation at Fort Worth, where the board met, furnished computing facilities to aid in the investigation.

I was very impressed by the effectiveness of this small group of experts, who knew exactly how to proceed with investigating the problems encountered in the accident. I thought how desirable it would be to have such a capable group assigned as my own engineers for conducting the projects in my division. Of course, such an arrangement would not be possible because the group consisted of the top men from each company or organization, each one of whom normally supervised a group of less experienced engineers.

The proceedings of this board were classified secret. As a result, I was not allowed to take out any notes or documents. My account of this investigation is therefore given entirely from memory and may be subject to errors. I will attempt, however, to give a correct general impression of the investigation.

The emphasis, of course, was on the control system of the airplane. This control system had been designed by the Eclipse-Pioneer Division of the Bendix Corporation and incorporated many novel features. An attempt was made to place the various subsystems, such as the air data system, the gyroscope and accelerometer unit, the amplifier and computer assembly, the engine controls, and the autopilot into an integrated system. The various systems all came together in the power control linkage assembly or PCLA. This complex mechanism, which fit into a space about 2-feet wide by 3-feet long, contained all the equipment for receiving the pilot's inputs and the autopilot signals and provided control feel, ratio changing, commands to the control surfaces, and various safety and monitoring provisions. Usually, such functions are distributed throughout the airplane, and the complexity of this tightly packed unit must have caused a nightmare for maintenance. In fact, as I recall, the unit was intended to be returned to the factory for maintenance. In accordance with the state of the art in that time period, the system used all hydraulic power controls and all analog computation.

As it turned out, the lateral-directional system soon came under suspicion as the cause of the accident. All the quantities sensed by the autopilot were fed to the control system by a control law that required changing gains as a function of flight condition. These gains were varied by servo-driven potentiometers, or pots, in a manner similar to that used in analog computers of the time. Investigation of the wreckage showed that one of the pots in the autopilot was against its stop, which indicated that the electric servo driving the pot had probably experienced a hard-over signal. Normally, it would be expected that such a failure would change the handling of the airplane somewhat, but would not cause a catastrophic failure. In this case, however, when the pot was driven against its stop, the wiper on the pot ran off the end of the resistance element and produced zero output. As a result, the gain of this particular feedback signal was zero. Stability calculations showed that with this feedback gain at zero and the others at their normal values, the airplane would have a highly unstable lateral, or Dutch roll oscillation. Furthermore, with this particular combination of gains, the normal control actions of the pilot in attempting to keep the wings level would further destabilize the oscillation. It was therefore concluded that when the failure occurred, the airplane went into a divergent lateral oscillation and reached sideslip angles sufficient to break off the tail.

The findings of the board were undoubtedly transmitted to the proper officials in the Air Force, and correction of the problem might involve simply limiting the travel of the servos driving the pots to avoid exceeding the ranges required in flight. This accident, however, illustrates the problems that caused great reluctance of airplane manufacturers to adopt electronic controls for airplanes and that required 20 years or more before commercial airplane manufacturers would use such devices in primary control systems, despite the improved capabilities and lower cost and weight of these systems.

For many years, the design of each new airplane had been supervised by a chief engineer, who was sufficiently familiar with all elements of airplane design and construction that he could understand the operation of these systems and could bring his long experience to bear in avoiding dangerous features. As soon as electronic control systems were introduced, the operation of these systems was beyond the knowledge of the chief engineer. In the case of the B-58, a company skilled in autopilot design was given responsibility for the design of the control system. The electrical and electronic engineers, however, did not have long experience with the safety problems of airplanes or of the possible catastrophic effects of a very minor failure. In my experience in flight research, I worked with airborne radar equipment that employed several hundred vacuum tubes. Usually two or more tubes would have to be replaced after a few flights, yet the designers of this system considered it to be highly reliable. Later, more extensive studies of reliability requirements for control systems were made. The basis of these studies were that no failure should occur, not in the lifetime of an airplane, but in the lifetime of the entire fleet of airplanes of a given design. This requirement amounts to a time between failures of the order of 10^9 flight hours. The meeting of such requirements is still a major problem, but with redundant systems and digital control equipment, these reliability goals can be met. Just the realization on the part of the designers of the importance of this degree of reliability represents a great advance in attaining the desired degree of safety.

Following the accident investigation, B-58 bombers continued to have crashes. Accident number 6 was a tail failure. Number 7 was an interesting case of an unanticipated problem resulting from the novel design of the flight control system. A feature of the longitudinal control system was that the airplane was automatically in trim at each value of airspeed. This feature is desirable on supersonic airplanes because large trim changes often

occur in going through the transonic speed range. Of course, this feature means that the airplane has no speed stability. Normally, a statically stable airplane without an autopilot can be trimmed to fly at a given airspeed and, because of the stable variation of pitching moment with angle of attack, it will tend to return to the angle of attack corresponding to the trim condition. The B-58 did not have this capability, but this problem was overcome by use of the autopilot system to automatically hold any desired value of airspeed.

One of the requirements in Gilruth's requirements for satisfactory handling qualities is that an airplane should require an increase in pull force on the control stick to reduce the speed to the stall. This feature is stated to provide a valuable form of stall warning. In the case of the B-58 in a flight in icing conditions, the airspeed head iced up so that the pilot had no correct instrument indication of airspeed. Also, a delta-wing airplane like the B-58 has a gradual stall with no buffeting until large values of angle of attack are reached. The result was that the B-58 gradually slowed down without the pilot's knowledge until it stalled and fell out of control. This experience shows that the provisions of Gilruth's flying qualities requirements should be kept in mind when devising new control systems. If the new systems fail to provide these features, some modifications should be made to provide an equal degree of safety.

Eventually, the problems of the B-58 were worked out and pilot error was decreased with the introduction of training versions of the airplane. In all, over 80 of these airplanes were manufactured. They never saw service in war because the expense of operating them was found to be very large when compared with that of subsonic bombers like the B-52. The development of the weapons pod was dropped, but the airplanes were used for high-altitude reconnaissance and provided valuable surveillance information in the Cuban missile crisis and in a volcanic eruption in Alaska.

Accidents of the Republic F-105 Thunderchief Airplane

A number of cases had occurred in which the Republic F-105 airplane, following a pull out from a dive-bombing run, experienced a large amplitude pitching oscillation. In one case at Nellis Air Force Base, three airplanes were performing rocket missions in a race track pattern. One of the airplanes made a pull up at 500 feet altitude, and a series of three longitudinal oscillations was observed with accelerations of $7g$ and $-1g$. The following airplane in the pattern, after pulling out at 300 feet altitude, experienced a similar oscillation. At the end of the third cycle, the airplane went to a large pitch attitude and crashed.

I made a trip with an instrumentation engineer, R. H. Sproull, to visit the Republic Aircraft Corporation at Farmingdale, Long Island, New York, and then to the Aeronautical Systems Division, Wright-Patterson Air Force Base, to discuss the accidents. Prior to the trip, the Republic Corporation had already made modifications to the control system and had run flight tests at Eglin Field to study the problem. The main problems facing the company were whether to unground the airplane and whether to place restrictions on maneuvers.

The stabilizer control system incorporated a pitch damper in series with the stabilizer actuator. To protect against pitch damper failure, the pitch damper was originally designed to cut off automatically if the acceleration exceeded $4g$ or $-1g$, followed by a pitch damper actuator deflection of greater than 90 percent full travel in the direction to extend the g value beyond these limits. The pitch damper could be caused to disengage, therefore, by a pull up exceeding $4g$, followed by a rapid push down which caused the damper to oppose the negative pitching velocity.

The airplane also had very light stick forces and low longitudinal stability in the conditions existing in dive pull outs. Such conditions are known to be favorable to pilot-induced oscillations. The control feel was considered desirable by pilots, however, and it was generally considered that the problem of oscillations could be avoided by pilot training. The Republic Corporation had realized that a pitch damper disconnect in the middle of a maneuver might trigger a pilot-induced oscillation. The pitch damper system had therefore been modified to prevent pitch damper disconnect except in case of an actual failure of the pitch damper. Though details of this modification are not available, the change involved installing a duplicate pitch gyroscope and a monitor circuit.

In examining the data from tests of the modified system, the NASA representatives noted some inconsistencies between the measured stick forces and the movement of the control stick. These data, if correct, could indicate a malfunction of the control system. This discovery caused some consternation among the Republic engineers, but it was soon realized that the data might be incorrect if up and down forces on the control stick could affect the strain gauge instrument that measured stick force. Such forces might be applied by the pilot in pull-up and push-down maneuvers to prevent his head from bouncing against the canopy. Tests were made that confirmed this source of error.

The final recommendation by the NASA representatives included removing the grounding restrictions on the airplane, but adding restrictions on dive angle and g value level during pull outs. This accident investigation reemphasized the knowledge that has been gained from numerous cases, namely, that stabilizing devices should not be disconnected as a safety measure, particularly in situations where the airplane is performing violent maneuvers.

Analytical Study of the F-84 Following Accident in High-Speed Flyby

Several airplanes have broken apart in the air during high-speed flybys. One of these was the Republic F-84 Thunderjet, an early unswept-wing jet airplane with tip tanks. Such an accident, which is tragic enough in any case, is doubly embarrassing to the company because it occurs during a demonstration of the airplane to high officials of the government or the Air Force. Movies of the crash seem to show that the airplane exploded in the air, but a frame-by-frame analysis of the movie shows that the airplane actually pulled up very rapidly to a high value of normal acceleration, about 8g, before the breakup occurred.

I was curious about the cause of this and other similar crashes. Though I was not associated with the accident investigation board, I made a brief analytical study of the stability characteristics that might contribute to a crash of this type. The study was never published, but it resulted in a memorandum for the Chief of Research dated October 11, 1948, with the title "Effect of Approach to the Wing Divergence Speed on the Longitudinal Stability of an Airplane."

Wing divergence is a phenomenon caused by the lift on a wing acting at a point ahead of the wing torsional axis. The torsional axis is defined as the axis location at which a lift force will produce no twist of the wing. Ordinarily, unswept wings have an aerodynamic center, or point of application of the lift force due to change in angle of attack, at about 25 percent of the chord, but this point would be still further forward with tip tanks. The torsional axis is usually at 35 to 45 percent of the chord. During a maneuver, the wing therefore twists in a direction to increase the lift and torsional moment above those of a rigid wing. If the structural restoring moment exceeds the destabilizing effect of the air forces, the wing behaves normally.

Beyond some value of dynamic pressure, however, the wing becomes unstable and diverges until the structural strength is exceeded.

The dynamic pressure for divergence is defined as the dynamic pressure at which the wing diverges when the wing root is rigidly fixed. Airplanes are flown so that this dynamic pressure is not exceeded in normal operations. The fuselage of an airplane in flight, however, does not provide a fixed restraint because of the change in angle of attack of the fuselage during a maneuver. Some effects on the motion of the airplane might therefore be expected at speeds below the divergence speed.

Equations for the longitudinal motion of the airplane were set up with the assumption that in addition to the usual degrees of freedom (vertical motion and angle of attack) a third degree of freedom representing wing torsion was included. The twist angle, of course, varies with the position along the wing span. To account for this effect, a technique known as the semirigid method, described in some British reports, was used (ref. 15.2). A variation of twist angle along the span was assumed, as well as a distribution of torsional moment of inertia. By use of a formulation of the equations of motion called Lagrange's equations, which integrates these effects, the torsional mode may be specified by a single variable representing, for example, the twist angle at the tip.

For a complete analysis of this problem, wing bending should also be included, but it was reasoned that wing bending would have less effect on the motion because wing bending does not directly affect the pitching moments acting on the airplane. Because of the lack of computational facilities at the time of this study, it was necessary to limit the number of variables. With the variables assumed, the resulting motion consisted of the short-period rigid-body motion and the torsional mode of the wing. With the assumption of constant airspeed, the analysis therefore required the solution of a fourth degree equation, for which as mentioned previously, solution techniques were available.

The airplane characteristics assumed did not represent any given airplane, but were representative of a high-speed fighter airplane. The wing divergence speed was assumed to correspond to Mach number of 0.9 at sea level. The center of gravity was assumed to be at 31 percent of the mean aerodynamic chord (MAC), which gave a stability margin in maneuvers at low speed of 2.5 percent MAC. The point of application of lift due to wing torsion was at 25 percent MAC. The results showed that when the wing torsion was taken into account, the airplane became longitudinally unstable in maneuvers above a Mach number at sea level of 0.61, and at a Mach number of 0.8 the time for the longitudinal motion to double amplitude was less than 1 second.

A point of interest is that most flight tests of high-speed airplanes are made at altitudes greater than 15,000 feet. Often, a high-speed low-altitude pass to demonstrate the airplane is the first time that conditions of maximum dynamic pressure have been encountered. This circumstance may account for the number of accidents that have occurred in such demonstrations.

Problems of the type studied were encountered on the Republic F-84 and the Northrop F-89 Scorpion, both of which had large tip tanks. The cure in the case of the F-84 was to put large triangular fins on the outside rear of the tanks, thereby adding a stabilizing effect to the torsional moments acting on the wing, and as an added benefit, reducing the induced drag of the airplane. This modification was installed on the F-84E and later models. The success of such empirical fixes probably accounts for the lack of interest in more detailed analytical studies of this problem.

A nonlinear structural phenomenon also was evident on many airplanes of this period that resulted in abrupt, unexpected incidents of both wing flutter or airplane instability. The wings of airplanes of this period where rela-

tively thick with thin-skinned monocoque construction. These wings showed adequate torsional stiffness when tested on the ground. When the wing was subject to high loads in flight, however, the skin wrinkled extensively, which resulted in a loss of torsional rigidity. In a high-g-value maneuver, the airplane could therefore be placed instantly in a very unstable condition that was not evident in flight preceding the maneuver. Fortunately, these problems have largely disappeared on modern airplanes that have thinner wings and thicker wing skin.

Accidents of the Beech Bonanza Involving In-Flight Structural Failures

Chapter 10 contained a discussion of the problem of airplane crashes resulting from spiral instability in flight under instrument conditions. This type of crash usually resulted from a pilot with insufficient experience in instrument flight becoming disoriented and entering a spiral dive. With early types of light airplanes, the airplane often was destroyed by crashing into the ground. If the airplane emerged from the clouds with sufficient altitude, the pilot could usually regain his orientation and recover. In the case of more modern personal airplanes with higher wing loading and lower drag than earlier types, excessive speed builds up rapidly in a spiral dive, which sometimes results in structural failure in the air regardless of the visibility conditions.

Accidents of this type involving in-flight structural failure have been encountered with all types of high-performance personal airplanes, but one type, the Beech Bonanza, had an accident rate much higher than other similar airplanes. This airplane, with a distinctive V-tail, was first introduced in 1947 (figure 15.5). The company stopped building the V-tail model in 1982 after 10,405 had been produced. By 1986 there had been

FIGURE 15.5. Beech B-35 Bonanza airplane as tested at Langley.

232 reported in-flight structural failures of these airplanes, which resulted in deaths of over 500 people. Already in 1950 the high accident rate of this airplane had become apparent. As a result, the Civil Aeronautics Administration (CAA) requested that the NACA conduct flight tests on the Bonanza to try to determine any unusual characteristics that could contribute to the accidents. This investigation was conducted by personnel in my branch and a report was published (ref. 15.3).

The flight tests of the Bonanza were conducted by NACA test pilots and, in accordance with safety procedures followed in all NACA flight tests, were limited to the placard values of airspeed and acceleration, which were an indicated airspeed of 200 miles per hour and a normal acceleration of 4g. Within these limits, no unusual or dangerous characteristics were observed. Special tests were made in which the pilot in a hooded cockpit was given control of the airplane in inverted and other unusual attitudes and was required to recover with the most elementary flight instruments, the so-called needle-ball-airspeed group. A skilled pilot encountered no difficulty in making these recoveries. The pilots did note that because of the clean design of the airplane, speed could build up beyond the placard limit rather rapidly if the airplane were nosed down.

Accident data showed that many of the in-flight breakups were encountered at values of airspeed well beyond the placard limit. Any such accidents are not considered the fault of the designer, because the pilot is never supposed to exceed the placard limit. Furthermore, the CAA design rules did not specify any design requirements for safety or handling qualities beyond the placard limits.

In studying the flight data, I observed that the curve of control force against dynamic pressure in straight flight had a downward curvature in the direction of increasing push forces with increasing speed. This plot, for a rigid airplane, should be a straight line. Various types of aeroelastic distortion can result in the plot deviating from a straight line. The Bonanza, in common with some other light planes, had a large negative setting of the stabilizer to allow the nose to be raised on take-off at a forward center-of-gravity location. This stabilizer setting resulted in a down-elevator deflection for trim in cruising or high-speed flight. Such an elevator deflection would cause the stabilizer to twist by an amount that increased directly with the dynamic pressure. In a V-tail design, the twist would be increased when compared with a conventional stabilizer because of the

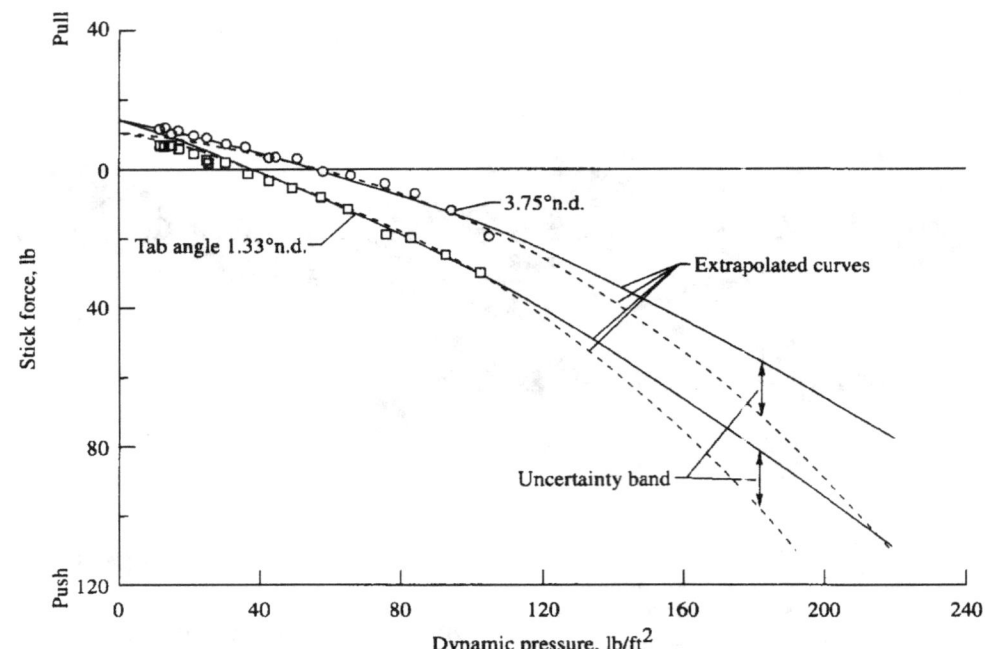

FIGURE 15.6. Measured and extrapolated stick force in straight flight as a function of dynamic pressure for Beech B-35 Bonanza airplane.

increased length of the tail panels. With a rigid stabilizer, the curve of control force against indicated airspeed would be a parabolic curve, which varied as the square of the speed, because the dynamic pressure varies as the square of the speed. The effect of the stabilizer twist would be to add to this curve an amount varying as the fourth power of the speed. The result of this addition would be to make the push force required for trim in $1g$ flight increase rather rapidly at high values of airspeed.

The plot showing experimental values of control force versus dynamic pressure for two values of trim speed is shown on figure 15.6. Because of the limitation of the flight data to 200 miles per hour, extrapolation of the data to higher values of airspeed is difficult. Nevertheless, two extrapolated curves are shown. From the upper extrapolated curve, the one less affected by stabilizer twist, two plots of control force versus indicated airspeed for various values of normal acceleration are shown in figures 15.7 and 15.8. The values of control force are lit-

tle affected by bank angle or pitch angle. These curves therefore apply during a spiral dive as well as in straight flight.

The conclusion from this analysis is that the airplane has a great tendency to pull up to higher values of normal acceleration if the airspeed increases beyond the placard speed in a spiral dive. For example, if the pilot holds zero control force, the acceleration at 280 miles per hour would be $6g$ with a trim speed of 145 miles per hour or $8.5g$ with a trim speed of 120 miles per hour. Normally, if the pilot becomes disoriented in a spiral dive, he will exert a pull force on the control wheel, further increasing the loads on the airplane. It is considered very unlikely that he would exert a push force to reduce the loads on the airplane. The results therefore indicate that structural failure might result because of the normal acceleration exceeding the design strength of the wing or of the tail. These results are based entirely on analytical extrapolation of the data obtained within the placard limits. No actual tests have been

FIGURE 15.7. Estimated stick force as a function of indicated airspeed and normal acceleration for Beech B-35 Bonanza airplane. Trim speed 120 miles per hour. Crosshatched area indicates limits of test data.

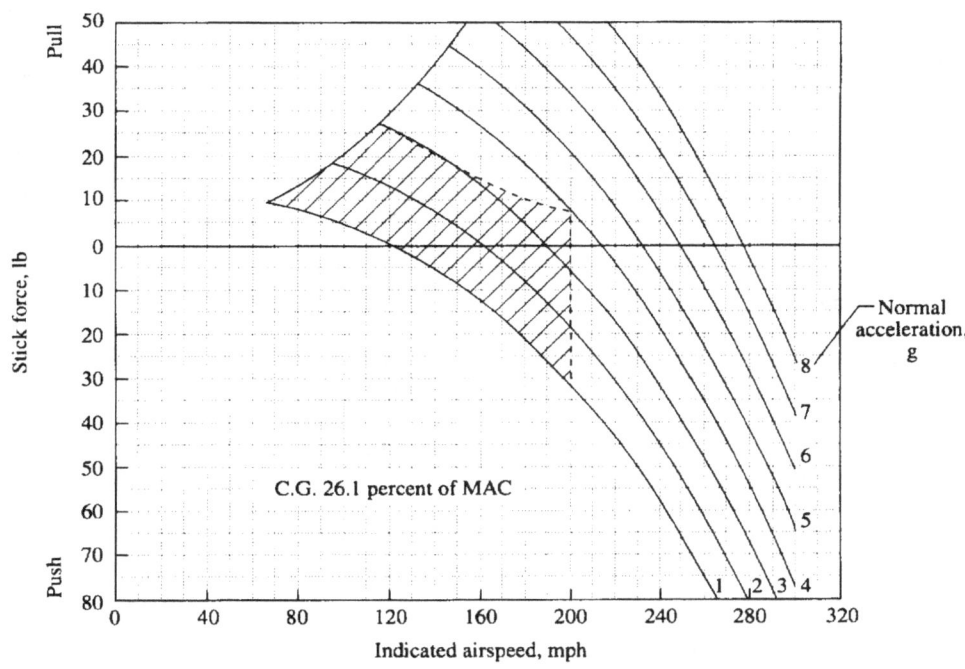

FIGURE 15.8. Estimated stick force as a function of indicated airspeed and normal acceleration for Beech B-35 Bonanza airplane. Trim speed 145 miles per hour. Crosshatched area indicates limits of test data.

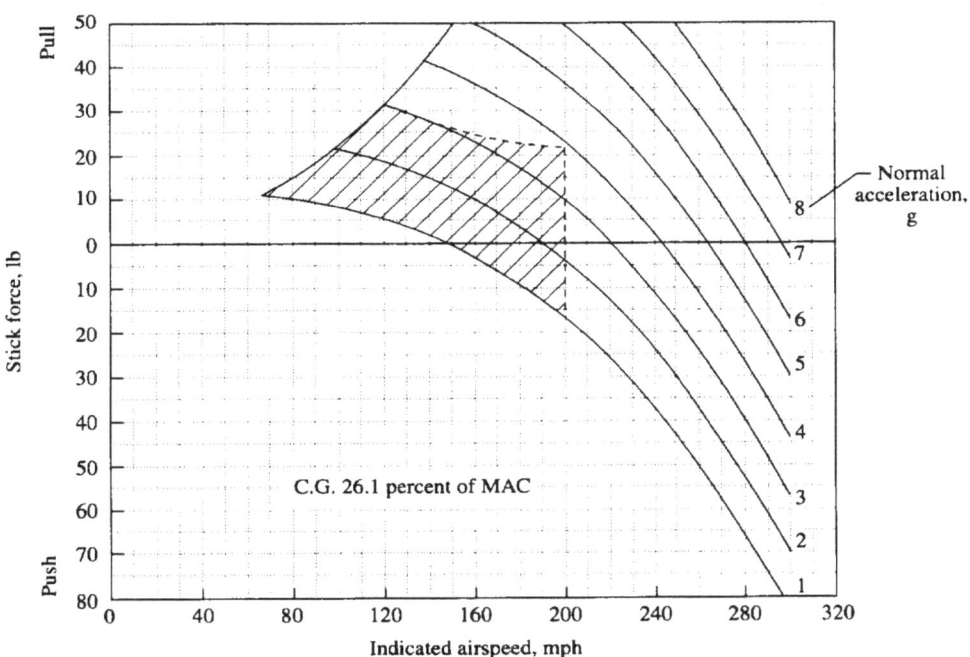

made to study these problems in the speed range beyond the placard limits.

One cause of the increased accident rate of the Beech Bonanza may therefore be the unusual tendency of the airplane to pull up in flight at high speed beyond the placard limit. Though no design requirements are specified for this regime of flight, a large improvement in safety might be attained with very little penalty in weight or performance by considering the handling qualities in this regime of flight. Most modern airplanes would not suffer catastrophic failure in a high-speed dive unless the acceleration increased beyond the structural limits.

At the time these studies were made, they were discussed with members of the Beech Corporation and of the CAA. No immediate action was taken, because of the philosophy that consideration of handling qualities beyond the placard limit was not required. In later years, models of the Bonanza with increased power and gross weight were produced, and some increases in the structural strength of the stabilizer were made by the Beech Corporation. Accidents continued, however, until in 1984, an extensive study was made by a task force at the Transporta-tion Systems Center of the U.S. Department of Transportation. The purpose of this study was to assess the facts associated with the controversy that had arisen regarding the design and certification process for the Bonanza. At this time, I transmitted the studies made by the NACA in 1950 to the director of the study. The results of the study by the Transportation Systems Center are contained in a comprehensive report (ref. 15.4).

Despite these extensive efforts, no change in design philosophy or certification procedure has been made. Though the final report of the Bonanza task force contains considerable study to find any case of failure within the placard limits, no mention is made of the possible failures beyond the placard limits, and no consideration is given to the idea of improving the characteristics in this range. Production of the V-tail Bonanza was discontinued in 1982 and later models of this airplane with conventional tail surfaces have shown much reduced accident rates. With the present lack of any effort to give consideration to characteristics beyond the placard limits, I believe that problems of this type may occur with future designs and that the lessons learned from the Bonanza should not be forgotten.

CHAPTER 16

NACA Flight Research on the Vought F8U-1 Airplane

Flight measurements of handling qualities of numerous airplanes had continued through the years following WW II at both the Langley and the Ames Laboratories, until about 60 airplanes had been tested by the mid 1950's. The flight research on the Vought F8U-1 (fig. 16.1), an early Navy supersonic fighter airplane, is singled out for special emphasis because it was the last high-speed airplane on which handling-qualities tests were conducted at Langley. In 1958, a decision was made by the NACA Headquarters to conduct all testing of high-speed airplanes at Edwards Air Force Base, California. This decision was based partly on considerations of safety because of the large population of the area surrounding Langley Field and partly to consolidate facilities to save money. From the standpoint of the Flight Research Division at Langley, the change represented a considerable loss of accumulated experience and expertise, as well as instrumentation and evaluation techniques that had been developed over the years. Soon after this decision was made, however, the appearance of Sputnik resulted in the conversion of the NACA to NASA. The dawn of the space age brought many new problems, which resulted in a decline in interest in aeronautical research. Many of the engineers who had worked in my branch on handling qualities left to join the Space Task Group that later moved to the Johnson Space Center in Houston, Texas, to form the nucleus of the

Manned Space Flight Program. Other engineers in my branch who remained at Langley were employed in simulation studies of aeronautical and space operations.

The flight tests of the Vought F8U-1 airplane were considered by the Langley Director, Dr. Floyd Thompson, as being especially important. As a result, he requested that frequent reports be written, for his attention, to summarize the progress on the job. I assigned Christopher C. Kraft, Jr., one of my most capable engineers, as project engineer. Donald C. Cheatham, a former Navy pilot, assisted in the many discussions with the Navy. William S. Aiken, Jr. of the Loads Division was assigned to aspects of the work involving measurement of structural loads.

The numerous progress reports were originally classified secret, but in 1981 after many years had passed, I had these reports declassified and collected them in a document (ref. 16.1). This collection of progress reports presents an interesting historical account of a typical handling-qualities investigation and shows how this work helped to improve the quality of Air Force and Navy airplanes during the period after WW II. This document was never published, but copies are available at the NASA Langley library.

Shortly after the beginning of the space program, Christopher C. Kraft, Jr. moved to the Johnson Space Center. He became a Flight

FIGURE 16.1. Vought F8U-1 airplane showing variable-incidence wing in raised position.

Director in the Apollo Program, and after the retirement of Dr. Robert R. Gilruth, became Director of the Johnson Space Center during the latter stages of the Apollo Program. During the development of the Space Shuttle, Donald C. Cheatham also moved to the Johnson Space Center, where he played a prominent role in the development of the Shuttle control system. William S. Aiken, Jr. moved to NASA Headquarters in Washington, where he held various positions, including director of the NASA Aeronautics Program.

Results of the Last NACA Flying Qualities Investigation

The progress reports on the F8U-1 investigation, written mainly by Christopher C. Kraft, Jr., give a good insight into the interaction between the Navy, the Vought Company, and the NACA in developing this fighter airplane. In addition, they catch the spirit of urgency attached to this study because of crashes that occurred in the early development of the airplane. They also record the inevitable delays and difficulties of flight research caused by lack of spare parts, problems with instrumentation, and weather.

The F8U-1 airplane was received at Langley in December 1956 as the result of a policy by the Navy to turn over to the NACA an early production model of each of its new fighter airplanes for flight evaluation and research. A novel feature of this airplane was a variable-incidence wing to place the fuselage in a more nearly level attitude for landing on a carrier. An early source of difficulty was inability to move the variable-incidence wing to the down and locked position after takeoff. This deficiency was responsible for the loss of at least one airplane and was a cause of frequent aborting of flight missions. The Aircraft Loads Branch was called into the study. Measurement of the forces in the wing-strut actuator during the retraction and extension cycle provided the information necessary for the company to design a wing-strut actuator with adequate power. In addition, operational sequences were determined to allow satisfactory use of the airplanes with the original strut that were already in service.

The longitudinal stability and control measurements of the airplane showed a decrease in stability at high values of acceleration in maneuvers. This deficiency was emphasized by a crash of an airplane during a flyby at the Vought plant. This crash, like that of the F-84 discussed previously, was a highly visible disaster as the airplane was being demonstrated to a group of Navy test pilots. The cause of the difficulty was determined by quantitative measurements of the deflections in various parts of the control system in accelerated maneuvers. Deflection of the fuselage under load was found to cause a movement of the stabilizer. An up load on the airplane caused a deflection of the stabilizer to a more negative incidence, which was a direction that tended to increase the upward load. The effect was therefore in a direction to destabilize the airplane. Once the cause of the problem was established, the cure was simple. Reversing the position of a link in the elevator control system caused the fuselage bending to stabilize the airplane rather than destabilize it. This modification was retrofitted to all airplanes. Prior to this modification, the low-altitude maneuverability of these airplanes was severely restricted for reasons of safety.

In addition to these important contributions to the safety of the airplane, the studies on the F8U-1 provided considerable experience in the general area of handling qualities of airplanes in supersonic flight. The airplane was one of the first to be fitted with roll and yaw dampers. The flight tests provided information of interest in the required stabilizing effects of these systems.

Another novel feature of the control system was a nonlinear linkage between the stick motion and the stabilizer deflection. The nonlinearity was such that the stabilizer motion was zero for small stick deflections about neutral, but increased with increasing slope as the stabilizer deflection increased. This feature was intended to offset the increased effectiveness of the stabilizer with increasing air speed. At high values of speed, small stabilizer deflections are usually used because they produce large loads on the airplane. The small sensitivity of the control stick at small deflections was intended to allow larger stick deflections in this regime. These larger stick deflections give the pilot more accurate control of acceleration. At low speeds, where the stabilizer effectiveness is decreased, the control-stick movement, and hence sensitivity, was increased. This feature appears logical. A problem arises, however, in that when the pilot has become accustomed to flying with gentle maneuvers for a period of time, he will be surprised by the large increase in control sensitivity in a large-amplitude maneuver. Some nonlinearity in the control-stick gearing may be beneficial, but the amount used on the F8U-1, which caused the stick sensitivity to approach zero at zero deflection, was found to be excessive. The results of the studies of the flying qualities of the F8U-1 are given in a memorandum report (ref. 16.2).

CHAPTER 17

A Couple of Bloopers!

I have previously called attention to the fact that engineers sometimes come up with bad ideas. I can't claim to be immune from this problem. Confession to a couple of projects that didn't turn out exactly as planned may therefore be appropriate. Fortunately, no great harm was done, and I learned some lessons from these experiences.

Improving on the Wright Brothers

When the Wright brothers first made their gliding experiments at Kitty Hawk, their biplane gliders had no vertical tails. They soon found that attempts to make turns by warping the wings were unsuccessful because the drag of the wing on the side twisted to increase the lift would cause the glider to swing around in the opposite direction from the desired turn. This phenomenon has since been called adverse aileron yaw. The Wright brothers then tried mounting rudders behind the wings that were linked to the wing warping controls so that they would deflect to offset the adverse yaw. The result was immediate success in producing good control in turns.

Later, the Wright brothers and most other airplane designers gave the pilot independent control of the rudders so that the human pilot could use the rudder as desired to offset the aileron yaw and could also use the rudder for other purposes such as ground control, landing in crosswinds, and spin recovery. Nevertheless, elimination of a separate control for the rudder remained a subject of interest because it was thought that correct use of the rudders was a difficult problem for novice pilots and that airplanes to be flown by the general public should be as simple to control as automobiles. Airplanes in which the rudder control was eliminated were called two-control airplanes. Perhaps the most notable examples of airplanes of this type were the early models of the Ercoupe designed by Fred Weick.

A problem with the direct linkage between the rudder and ailerons as used by the Wright brothers is that the amount of rudder to be used varies with the airspeed. This problem was not serious for the Wright brothers because the speed range between the stall speed and the maximum speed was very small. This variation arises because the adverse yaw of the ailerons increases directly with the lift coefficient. As a result, much more rudder deflection is required to make properly coordinated turns at low speed than at high speed. Human pilots can approximate the required variation by applying the same force to the rudder pedals for a given aileron deflection at any airspeed. On manually controlled airplanes, the rudder deflection produced then varies inversely as the square of the speed, exactly the variation required to

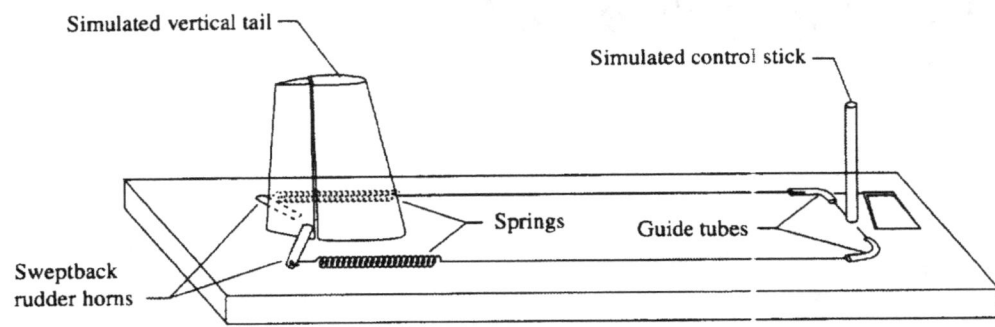

FIGURE 17.1. Sketch of model used to illustrate the use of sweptback rudder horns to offset centering tendency of control springs.

give a yawing moment that increases directly with the lift coefficient.

One method to cause the amount of rudder deflection to decrease as the airspeed increased would be to operate the rudder from the aileron control system through a spring. The hinge moment applied to the rudder would then be independent of airspeed so long as the rudder was in neutral. When the rudder moved, however, the spring force would fall off due to the tension on the spring being released. As a result, the rudder deflection produced in low-speed flight would be less than that required, whereas in high-speed flight it would be more nearly equal to that required.

To avoid this problem, the restoring moment applied to the rudder by the interconnecting springs should be offset by an unstable spring moment that would tend to deflect the rudder away from neutral. I reasoned that a simple way to provide this effect would be to sweep the rudder horns back. I had occasionally used this technique in the past to vary a restoring moment from stable to unstable. I built a small model to demonstrate the principle. A sketch of this model is shown in figure 17.1. In this model, a small control stick was connected to threads that ran through pieces of curved tubing to act as guides. Then springs were attached to the ends of the threads and attached to the control horns on the rudder. By sweeping the horns back the correct amount, the rudder would stay in any position, which indicated that the restoring moment of the springs had been offset by the

unstable moment produced from the spring tension being applied aft of the hinge line.

The model worked perfectly, and it was therefore decided to try the system in a small low-wing trainer that was available at the hangar, a Fairchild PT-19. A photograph of this airplane is shown in figure 17.2. The installation was very simple and required only that new rudder horns be built and that springs be attached between the control cables and the rudder horns. In addition, control of the rudder was transferred from the rudder pedals to the aileron system. A sketch of the control system is given in figure 17.3.

When the system was installed, a disconcerting effect was found. With the airplane on the ground, the control stick would not stay in neutral, but would immediately slam rather violently to one side of the cockpit or the other. The unstable moments applied to the rudder destabilized the entire control system! This effect had not been noted on my model because the friction on the threads in the guide tubes had been enough to keep the control stick from moving.

After some consideration, additional springs were attached to the control stick to offset the unstable spring moments, as shown in figure 17.3. Then the airplane was considered safe to fly. Flight tests showed that the rudder coordination in turns was satisfactory, but the aileron control forces were excessive because the aileron control stick had to supply the force to deflect the rudder as well as the ailerons. Because the rudder

FIGURE 17.2. Fairchild
PT-19 airplane used for
two-control experiments.

FIGURE 17.3. Sketch of
modified directional
control system.

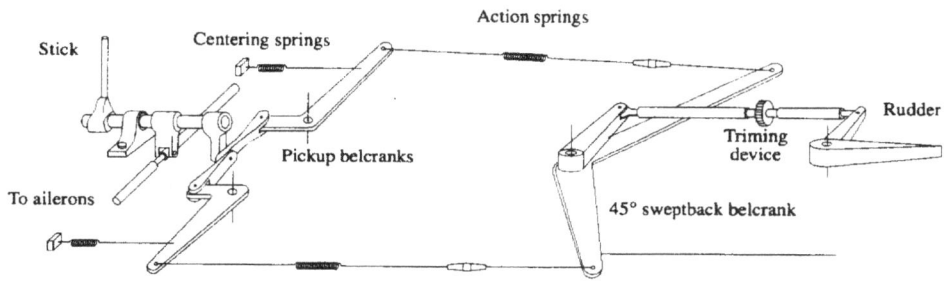

had been designed to operate with rudder pedals, to which the pilot can apply considerably more force with his legs than he can apply to the control stick with his arms, the rudder hinge moments increased the lateral control forces by a large amount. This problem could have been alleviated by redesigning the rudder to be more closely balanced aerodynamically, but the work involved in making this modification was not considered worth the effort.

A brief theoretical analysis showed that despite the provision of centering springs on the control stick, the control forces on the ground are unstable about neutral over a certain range of deflection. The reason for this effect is that the rudder, lacking any restoring tendency, goes to full deflection as soon as the control stick is moved a small amount. The springs in the rudder system then apply a force to the control stick causing it to move

away from neutral until the springs on the control stick balance this decentering force. As the airspeed increases during the takeoff run, this instability disappears.

A draft of a report on this project was prepared by Stanley Faber, an engineer in my branch, but the report was never published. Figures 17.2 and 17.3 are taken from this report.

As a result of this experiment, I learned that my crude models could not always be trusted to give the right guidance for full-scale design. The program also recalled the reasons that the Wright brothers had abandoned the two-control arrangement; namely, the other uses for the rudder (primarily, landing in crosswinds) that make it inadvisable to eliminate this control. Fred Weick retained the two-control arrangement on the Ercoupe airplanes that he produced, but when another manufacturer resumed production of the

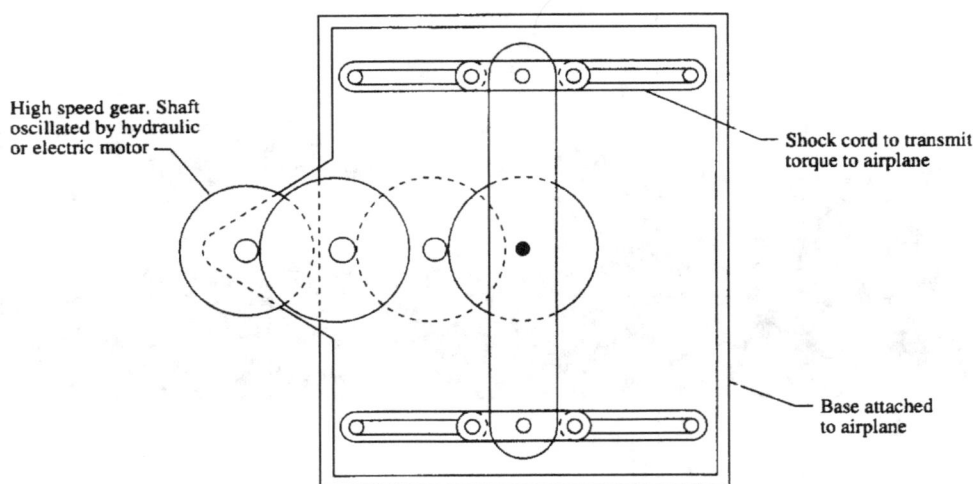

High speed gear. Shaft oscillated by hydraulic or electric motor —

Shock cord to transmit torque to airplane

Base attached to airplane

Ercoupe, the control system was changed back to a normal three-control system.

Relearning the Laws of Angular Momentum

A frequent objective of flight research is to determine the stability derivatives of an airplane; that is, such quantities as directional stability and damping in yaw. This procedure could be greatly simplified if some means were available to apply external forces or moments to the airplane in flight. In the early days of the NACA, the damping in roll of a Curtiss Jenny was measured by dropping sandbags of known weight from the wing tips and measuring the response of the airplane. This sort of procedure is not practical on modern, high-speed airplanes, however. Most research on measuring these characteristics has been conducted by measuring the response of the airplane following a displacement of the controls and using a mathematical process to fit the response with a set of derivatives that produce motion closely approximating the measured response.

Nevertheless, an idea occurred to me to build a device that would be capable of applying oscillating moments of varying frequencies to an airplane in flight. A sketch of this con-

cept is given in figure 17.4. Basically, the device consisted of a small flywheel that was spun back and forth at varying frequencies by an electric motor. Then the torque produced by the flywheel was magnified by running it through a gear train with a large reduction ratio. For example, a 10 to 1 reduction would multiply the torque by a factor of 10. If four stages of this gearing were used, with larger and stronger gears and shafting on each stage, the torque could be multiplied by a factor of 10,000. It appeared that the torque could be made large enough to actually oscillate the whole airplane in flight.

A device of this type was built and tried in a fighter airplane. On returning from the flight, the pilot said that he hadn't been able to detect any disturbance to the airplane.

Obviously, something was wrong. It didn't take much thought to realize that the torque applied to the airplane had to be equal to the change of angular momentum of the small flywheel and gearing, a rather negligible amount. The larger torques involved in the gear train were offset by reaction torques in the framework that held the gears, so they had no external effect. My face was rather red when I explained this situation to the pilots and the engineers involved.

CHAPTER 18

Administration of the Langley Research Center

In my career at Langley, my primary interest was in the technical aspects of my work. I paid little attention to how the center was directed or where the money came from to perform research. As a result, I am less capable of describing these problems of the center management than others who were involved in these aspects of research. Nevertheless, in the course of my work, I did learn something of the management of the center and was well aware of the importance of proper management. I will therefore attempt to summarize some of my impressions of the administration of the center during the years when the work was done under the NACA. Major changes came about with the advent of the space program and the change of the organization to NASA.

Organization of the Center

When I first came to work at the Langley Memorial Aeronautical Laboratory, I knew very little about its administration. Gradually, through bulletins that were circulated at rare intervals, I became aware of the general organization. In figure 18.1 is shown an organization chart for the Research Department in 1944. The director was called the Engineer in Charge, a title changed to Director a few years later. The Engineer in Charge was Dr. Henry J. E. Reid, who had started his

career about 1925 in instrument design, but soon went into administrative work. The second in command was Dr. Floyd Thompson, who was first called Chief of Research and later Associate Director. He was responsible for direction of the research programs at the center. Other departments at the center had to do with technical services and personnel. Most of the research divisions were broken into sections. A few had branches that, in turn, consisted or two or more sections.

Most of the eight research divisions shown in figure 18.1 were associated with a particular research facility or group of facilities. As a result, the divisions were in different locations. To keep all groups informed of the work that was going on, Dr. Thompson in about 1942, set up a program consisting of monthly meetings each hosted by a different research division that would present talks on important projects the group was working on. These meetings were first called Section Head Meetings because section heads usually presented the talks. Later they were called Department Meetings because all the material presented was from the Research Department. These meetings were held after working hours, but Thompson strongly recommended that all section heads or higher ranked supervisory employees should attend and most of them did. Naturally, each of the engineers presenting a talk before all the supervisory personnel at the center had a great incentive to do as good a job as

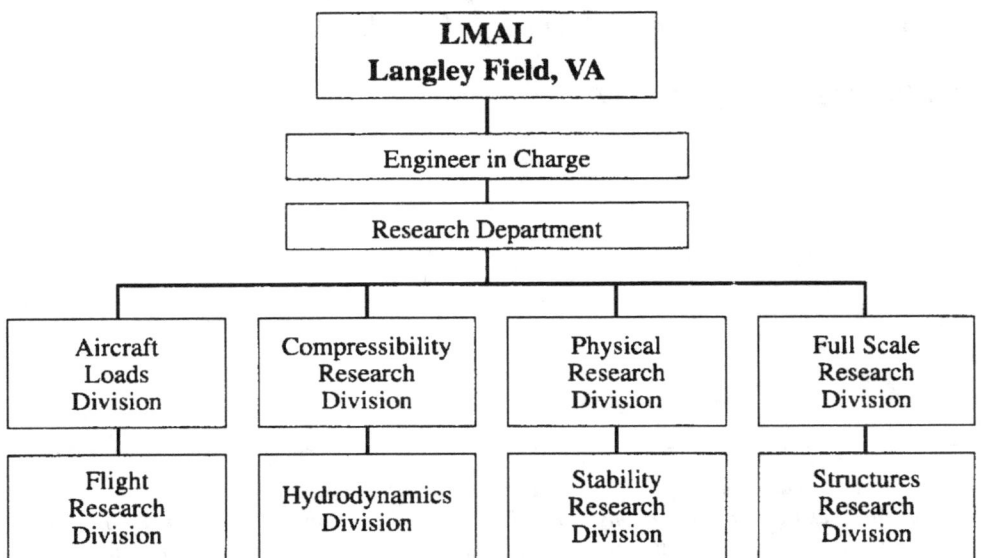

FIGURE 18.1.
Organization chart for
the Research
Department of the
Langley Memorial
Aeronautical Laboratory
in 1944.

possible. The talks were therefore very well prepared and usually were of great value in helping coordinate the work of the center.

NACA Reports

The main products of the work at Langley were reports written on the various research projects. The least formal type of report was called Preliminary Data, which was sent out to the military services or other agencies as rapidly as possible with little review. Later these reports were reviewed and put out as more finished documents of various types depending on the importance and security classification. These included RB's (Restricted Bulletins), MR's (Memorandum Reports), and CMR's (Confidential Memorandum Reports). The standard NACA publications existing before the war were TM's (Technical Memoranda), TN's (Technical Notes), and TR's (Technical Reports). These reports were unclassified and were always reviewed by a formal editorial committee of engineers and then by the Editorial Office, a group of ladies skilled in English grammar and the NACA rules for style and form.

The Technical Report was considered to contain the most important technical data or to have the most lasting interest. Technical Notes had less of these attributes and Technical Memoranda were usually either translations of foreign reports or reprints of important reports from other agencies. After the war, many reports were declassified and republished as Wartime Reports.

Great emphasis was placed on technical correctness as well as on clarity of presentation and logical arrangement. The editorial committee usually consisted of at least two engineers from the writer's division (one of whom was usually the chairman of the committee), one or more from other divisions, and one who was not working in the same field of work. The latter person, usually referred to as the "mean intelligence" helped to insure that unfamiliar theories and concepts were explained in sufficient detail to be understood by persons not working in the specialty of the writer.

My first assignment to an editorial committee was in the mean intelligence category. I was on a committee for a report written by Dr. Eugene Lundquist, Chief of the Structures Division, on theories of buckling of

FIGURE 18.2. Instructions to the members of an editorial committee.

COPY FOR _____

RESEARCH EDITORIAL COMMITTEE

	Name		Division	Branch	Section
1.	_____ (Chairman)		_____	_____	_____
2.	_____		_____	_____	_____
3.	_____		_____	_____	_____
4.	_____		_____	_____	_____
5.	_____		_____	_____	_____

Date _____   Time _____ a.m.   Room _____

NOTE TO AUTHOR: Collect all copies of report at end of editorial meeting.

AUTHOR:

TITLE OF PAPER:

RECOMMENDED FOR _____ R.A. _____ J.O. _____

TO THE COMMITTEE

The editorial committee to which you have been appointed constitutes a body chosen in accordance with procedures established by the Associate Director of the Langley Research Center to conduct a responsible technical editing of the attached paper for this Research Center.

You are therefore requested to review thoroughly the attached paper, including abstract and indexing, for technical accuracy, soundness, clarity, and general suitability for publication. The main editing should be done outside the meeting and you should come to the scheduled meeting with your points of criticisms and suggestions clearly in mind, preferably written.

TO THE CHAIRMAN

It is your responsibility to see that the committee performs the functions specified above and that the author revises the paper to the satisfaction of all concerned, if possible. If any important recommendations are not accepted by the author, Item 8 on the blue routing card (see card) is to be answered "No," and you are required to submit a memorandum to the Associate Director explaining the differences of opinion. You are also responsible for reviewing any changes made as a result of any comments received from Ames and Lewis to insure the coordination of such changes with recommendations of Langley editorial committee.

Your notation and initials on the routing card on Item 8, and initials of author's division chief on Item 9, are an indication to the Associate Director that the paper is now satisfactory for publication.

TO THE AUTHOR

After the editorial committee meeting, revise one copy of the paper, according to editorial committee recommendations and any Ames and Lewis comments received, to the satisfaction of yourself and your section and branch head. Initial Item 6, obtain initials of your section and branch head in Item 7, and forward report to the chairman of the editorial committee, Item 8 on the routing card. (See TO THE CHAIRMAN notes above.)

It is your responsibility to see that all other copies of the report are collected from the members of the editorial committee and destroyed so that there will be no possibility of an incorrect copy being used or referred to at any time. (See NOTE TO AUTHOR at top of page.)

L-96F353   NASA - Langley Field, Va.

thin shell structures. In those days, division chiefs actually took part in the research of their divisions and had time to write technical reports.

During my career, I served on many editorial committees, and each of my formal reports was the subject of a review by such a committee. The instructions for the editorial committee were given in a document reproduced as figure 18.2. This document was one of the longest lasting sets of instructions used at the center. It did not change in over 40 years. In 1980, the form was revised slightly to incorporate such modern ideas as use of "Chairperson" instead of "Chairman" and to account for some changes in the titles of officials in the chain of approval.

In general, editorial committees served their purpose well. In some instances, however, a person on the committee would try to incorporate ideas that involved a change in organization of the report rather than any change in the material presented. Such comments served to increase the time of report preparation, sometimes unduly. I have heard some engineers remark, "editorial committees are fine for everyone's reports—except mine!" The editorial committee serves as another means to keep groups informed of each other's work. One problem encountered more frequently in recent years is that the derivations or calculations involved are so complex that no one except the author is in a position to check the results. This problem is particularly apparent when most of the results are calculated by high-speed computers.

Approval and Support of Research

During the entire time that I worked in the Flight Research Division, from 1940 to almost 1960, I never had to worry or even think about the money required to do a job. The projects described up to now in this technical autobiography make it clear that much of the work was on ideas that I had proposed or that I directed people in my section to work on. To obtain approval for most projects, it was necessary only to have a talk with or write a memorandum for the Section Head. He, in turn, informed the Division Chief of the work he wanted to have done.

Evidently, the flight research work conducted at Langley was quite expensive, but I did not know the process for obtaining the necessary funds. The money for salaries for engineers, pilots, and mechanics must have been approved by Congress independently of the funding for research equipment, and the activities of these people were largely left up to the Division Chief. Though most of the airplanes under test were furnished by the military services, some were owned by the

NACA. The cost of fuel, spare parts, and equipment must have likewise come from another fund approved by Congress in the overall NASA budget. When a project was approved, it was necessary only to write a brief statement called a Job Order that described the objectives of the study and the estimated time for completion.

Keeping track of the progress or completion of a project was equally informal. When I became a Section Head, I kept a list of the R. A. (Research Appropriation number) and the J. O. (Job Order number) for each project under my supervision, along with a title of the job. When the job was completed or a report published, a line was drawn through that entry. Information that would be of interest today, such as engineers assigned and dates, were unfortunately not included. About 200 jobs are shown on the list over about a five-year period. Of course, formal lists were kept by the Langley management and at NACA Headquarters, but I never saw these lists.

This lack of concern with administrative matters was a very favorable feature of the research environment for the engineers who were devoting their energies to solving engineering problems. I have found that steady concentration on any difficult problem is necessary to make progress in its solution. Any distraction to other activities makes progress very difficult because the engineer must spend as much time in refamiliarizing himself with the work done to date as he did in doing it originally.

My main distraction of an administrative nature was writing raise recommendations for engineers when time arrived for their promotion. This was a continuing process, because of the large influx of personnel during the war years, which increased the total employment at the center from about 700 in 1940 to over 3000 by 1944. Although the promotion of satisfactory employees at regular intervals was a somewhat perfunctory event, a raise recommendation was nevertheless required in each case to make it clear

FIGURE 18.3. Certificate of appointment to the NACA Research Advisory Committee on Control, Guidance, and Navigation.

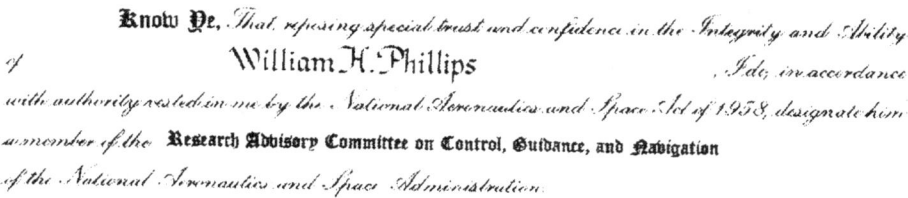

National Aeronautics and Space Administration

To all who shall see these Presents, Greeting:

Know Ye, That reposing special trust and confidence in the Integrity and Ability of William H. Phillips, I do, in accordance with authority vested in me by the National Aeronautics and Space Act of 1958, designate him as member of the **Research Advisory Committee on Control, Guidance, and Navigation** of the National Aeronautics and Space Administration.

In testimony whereof, I have subscribed my signature and caused the seal of the National Aeronautics and Space Administration to be hereunto affixed at the City of Washington, this **first** day of **July**, 1963.

Raymond L. Bisplinghoff
Associate Administrator for
Advanced Research and Technology

James E. Webb
Administrator

that the promotion was justified. Promotion of outstanding employees at a rate faster than normal required very strong and detailed recommendations.

The Committee Organization of the NACA

The initials NACA stand for National Advisory Committee for Aeronautics. Indeed, the technical governing body of the organization was this committee, often referred to as the Main Committee of the NACA. This committee was appointed by and reported to the President of the United States and its members served without compensation. Its members included representatives of the Air Force, Navy, Weather Bureau, Smithsonian Institution, and Bureau of Standards, as well as prominent aeronautical experts from industry and from universities. The NACA was an independent government organization and reported directly to the President. Under the Main Committee were committees representing the main disciplines of aeronautics, such as aerodynamics and structures. By the time I had worked a few years, these committees each had several subcommittees. I was appointed to the NACA subcommittee on Stability and Control, a subcommittee of the Aerodynamics Committee. Later, after 1958, when the NACA became NASA, the name subcommittee was dropped and the group became the NASA Research Advisory Committee on Control, Guidance, and Navigation. A certificate of appointment is shown as figure 18.3.

The subcommittee met four times a year, and its membership included many prominent stability and control experts from industry and from the universities. I frequently gave reports to the subcommittee on the work being conducted at Langley and on proposed research projects. It was my duty to carry back to Langley recommendations for research made by the subcommittee.

The committee structure had advantages and disadvantages. In theory, Langley and the other centers of the NACA were kept abreast of the needs of industry, and any promising new ideas were brought to light so that they could be studied with the facilities at the Research Centers. In return, the industry representatives were given the latest research results that might help them with the design of new airplanes.

In practice, the committees did not always produce the benefits desired. Industry representatives, generally representing large companies with their own extensive research facilities, did not readily reveal their latest ideas or results to their competitors on the committee. In addition, when the recommendations of the committee were taken back to Langley, a frequent response was that those items were already in the research program. Generally, university professors provided a more valuable interchange of information. On the whole, I considered the committee operation beneficial, but it was done at the expense of a lot of work and time that could have been used in carrying out the research program. After some years, when the NASA Headquarters organization had grown to the point that each discipline had its own group of administrators, the Research Advisory Committees were discontinued.

Formal NACA Conferences

A long-remembered technique used by the Langley Research Center to transmit its research results to industry was the Annual Inspection. As the name implies, once a year representatives of the industry and military services were invited to hear a series of presentations, usually one at each research facility. The event usually took two days and was noted for the efficient preparations that were made for transporting the groups, the impressive well-rehearsed talks by young engineers, and the live demonstration of facilities and equipment. When the Ames and the

Lewis Centers came into existence, the inspection was rotated among the centers, so that it came up once every three years.

In addition to the annual inspection, different disciplines such as Aerodynamics and Stability and Control presented conferences for the industry whenever they thought that enough new research information was available. Each discipline had a Langley committee, usually meeting monthly, at which new research results were presented and discussions were held on proposed research. These committees usually paralleled the NACA subcommittees. I was a member of the Stability and Control Committee, which was headed at first by T. Aubrey Harris, head of the Stability Division and later by Philip Donely, who was then head of the Flight Research Division. These leaders proposed and organized the conferences when they thought they were appropriate. In general, all the centers were invited to present papers at a conference.

The first such conference that I participated in was the Conference on Personal Airplane Research, which was held in 1946. This conference was intended to bring the personal airplane industry up to date on all the advancements made during the war that might be of benefit to them. At that time, there was great optimism that the personal airplane industry would enjoy a great expansion after the war, with all the military pilots returning to peacetime life. This expansion never materialized, however. Many of the advancements also made the airplanes more complex and expensive, which restricted their use to a specialized group who were either wealthy or who could justify the use of the airplane for business purposes.

Many of the later conferences, however, were favorably received because of the exciting nature of new results in transonic and supersonic aerodynamics. The bound volumes of papers presented at these conferences now form good summaries of research in this period of rapid progress in airplane performance.

Preparation of NACA and NASA Talks

The conference talks were carefully rehearsed and re-rehearsed, with the reviewers consisting of Branch Heads and Division Chiefs at the center. These officials had had long experience in this duty, in a tradition started by Frederick H. Norton, the first professional employee at Langley, and insisted on the figures being clear, legible, and uncluttered and on the talk presenting ideas clearly with a minimum of complex mathematics. This emphasis on clear presentation set an example for industry, the beneficial effects of which were soon evident in talks presented at professional societies and other conferences. The principles governing this approach are presented in a popular paper by Dr. Samuel Katzoff, entitled *Clarity in Technical Reporting* (ref. 18.1). I did my part in spreading the word in a lecture on report writing that was given to all new engineers, starting about 1960. The text of the lecture is given in a paper entitled "NASA Reports–A Discussion for Prospective Authors." Both these papers were widely circulated at the NACA and NASA centers.

I fear that the emphasis on clear presentations has largely disappeared with the retirement of the early Division Chiefs and with the advent of computer-generated figures, which usually have small print and illegible scales. Many such figures are prepared with fancy colors and designs, but are very unsuitable for transmitting technical ideas.

Meetings of Professional Societies

Other meetings that were very important for the dissemination of new research results were meetings of professional societies, primarily the Institute of Aeronautical Sciences (IAS), later called the American Institute of Aeronautics and Astronautics (AIAA). The meeting of greatest interest to me was the annual meeting of the IAS, which was held each January in the old Hotel Astor on Times Square in New York City. In a Victorian Ballroom decorated with pilasters capped by seminude plaster figures of smiling ladies, the most noted aeronautical scientists and engineers presented talks on the latest technical developments. In those days, there were no simultaneous sessions, so each talk was heard by the entire audience. I remember, for example, talks by W. Bailey Oswald of the Douglas Company on the advantages of four-engine airplanes, a presentation by Dr. James Lighthill on noise produced by jets, and a long and highly mathematical presentation by an unknown young Chinese scholar on boundary-layer oscillations. The latter talk is chiefly memorable for the merciless verbal lashing given the speaker by Max M. Munk, the famous early NACA theorist, who claimed that the speaker had "emasculated" the equations by omitting the nonlinear terms. I did not feel sorry for the speaker; he had violated all the rules of making a clear presentation.

To further emphasize the quality of the talks in the early meetings, I will list a few of the speakers given in the program for the 14th Annual meeting in January 1946

- I. L. Ashkenas, Northrop Aircraft

- Allen E. Puckett, Jet Propulsion Laboratory, California Institute of Technology

- George S. Schairer, Boeing Aircraft Co.

- Edmund V. Laitone, Curtiss Wright Corp.

- Hans Reissner, Polytechnic Institute of Brooklyn

- F. M. Rogallo, Langley Memorial Aeronautical Laboratory

- E. E. Aldrin, Colonel, Air Corps, Wright Field (father of astronaut Buzz Aldrin)

All of these speakers were known to me, either personally or by reputation, at the time of the meeting, and all are known today for their many additional contributions to the science of aeronautics.

Professional society meetings today do not compare with those early conferences in their value to a young engineer. The meetings have many simultaneous sessions with hundreds of papers, so that it is impossible to predict beforehand which papers will be best to listen to. Most of the papers are the results of thesis work by graduate students, who usually are trying to impress the audience with the complication of their studies. Some papers, of course, are valuable, but the chance of hearing one is rather small.

The American Physical Society meetings were usually held in New York at Columbia University during or immediately after the AIAA meetings. Sometimes I would go to hear a few sessions in that meeting also to acquire a slight acquaintance with such concepts as symmetry and charm in nuclear physics.

Meetings with Visitors

In addition to meetings of professional societies, there were many meetings held at Langley, either in the Flight Research Division or with center-wide groups, to discuss specific problems. Of special note are delegations from France and England in the postwar years. It is evident from my notes that much time was spent in such meetings and discussions and that outside organizations had a strong interest in the work conducted at the center.

CHAPTER 19

Dawn of the Space Age

I have never been a fan of science fiction. Model airplanes kept me abreast of the latest developments in aviation. In high school, 1935–1939, my doodlings showed sketches of highly streamlined airplanes, most of which were well ahead of military aircraft of the time. All of these designs, however, I knew to be practical and based on known technology.

During the 1950's, the Air Force was engaged in secret studies of orbiting spacecraft, as revealed in history books such as *Spaceflight Revolution* (ref. 19.1). My first inkling that flight in space was being considered at Langley was a talk at a department meeting, probably given by William J. O'Sullivan, Jr. He, along with Clinton E. Brown and Charles H. Zimmerman, had made comparisons of the efficiency of flight outside the atmosphere with flight in the atmosphere for long-distance travel. These studies showed that launching a ballistic rocket that would travel halfway around the world would require as much fuel as the flight of an airplane with a lift-drag ratio of 2 or 3. Of course, as the distance approached a complete orbit, the ballistic vehicle would continue to fly increasing distances with no further fuel expenditure so that the relative efficiency increased. Nevertheless, because halfway around the world was the greatest distance anyone would wish to travel and because supersonic airplanes had a lift-drag ratio of at least 6, I concluded that ballistic

travel was not very promising. The difference becomes much more striking when the type of propulsion is considered. An air-breathing engine such as that used in airplanes does not have to carry the oxygen required for combustion, whereas a rocket motor used for ballistic vehicles must carry both fuel and oxidizer. The weight of the oxidizer is several times the weight of the fuel.

I continued my research on airplanes and was as surprised as anyone when the newspapers announced the flight of Sputnik, a small experimental satellite launched by the Russians. This event resulted in great public interest and many attempts by amateurs to listen to the beep-beep signal as the satellite passed overhead. The public outcry was nothing, however, compared to that a couple of months later when the Russians orbited the Sputnik II, a two-ton satellite carrying a live dog. I knew that the weight of the object placed in orbit was only a very small fraction of the weight of the rocket required to launch it and was therefore very aware of the strides the Russians must have made in rocket propulsion.

My son, 9 years old at that time, kept a scrapbook as a school project in which he pasted many pictures of rockets and space ships, as well as clippings from newspaper articles related to the space program. A quotation from one article shows the concern felt about the Russian accomplishments (ref. 19.2)

New York, Jan. 6, (AP) - Malcomn P. McNair, widely known economist, said today the Russians in a very real sense already have won the cold war.

Sputnik has blown up our fool's paradise.

It is difficult to find words or tones of sufficient gravity to present what I am convinced is the true situation. I do not think we have as yet begun to realize the full picture partly because our leaders have been afraid to give us the facts.

In a very real sense, the Soviets have already won because it is now demonstrated that we cannot match Russian progress toward specific objectives of knowledge, power, and achievement without voluntarily submitting ourselves to a substantial degree of purposive direction in our lives, direction of investment, direction of manpower, direction of education.

The rapidly moving changes in the national space program, as well documented in *Spaceflight Revolution*, resulted on October 1, 1958, in the passing of the National Space Act and the formation of NASA, the National Aeronautics and Space Administration, from the former NACA centers, as well as the Army and Navy installations involved in space flight. I remember the comment made by Thomas J. Voglewede, head of the Performance Branch of the Flight Research Division, when he came to work that morning. He said, "The NACA is dead, long live the NASA." No doubt he pronounced the NACA and NASA as individual letters, as had been customary with the NACA employees. While this interpretation is still customary when referring to the NACA, the pronunciation of NASA as a single word became common soon after that and is now generally regarded as the correct designation for the organization.

The immediate result of this announcement was that every group immediately considered how its knowledge and expertise could be applied to aid the space effort. In the case of my Branch, I considered that studies of the stability and control of spacecraft would now be emphasized and that the portion of the space flight in the atmosphere during launch and entry would be fruitful subjects for research.

During this same period, another event occurred that had a pronounced effect on the work of the Flight Research Division. A Headquarters edict, published in 1958, stated that no further testing of high-speed air-

planes would be done at Langley. All future flight research on airplanes of this type was to be done at the NACA Flight Research Center at Edwards Air Force Base in California (now called the Dryden Flight Research Center). The reasons for this change were primarily to save money by consolidating research efforts and secondarily for safety reasons because of the perceived dangers of flight of experimental aircraft in a populated area.

The sudden emphasis on space flight could be considered fortunate as a highly interesting and popular change of activity to occupy the personnel who had previously conducted flight research on airplanes. Actually, the reason for optimism was more fundamental. Aeronautical research had reached a plateau around 1958. Many of the research contributions of Langley and the other NACA centers had reached fruition in the design of advanced airplanes. These airplanes included jet bombers and transports, supersonic fighter airplanes, and the British Concorde and the Russian TU-144 supersonic transports. No really revolutionary steps forward for atmospheric aircraft were envisioned at that time, nor have occurred in the ensuing years.

The change in emphasis in research, as well as the subsequent change in organization of the NACA Headquarters and of the Research Centers brought about by the space program, makes this a suitable time to end this book. Much research that was conducted during the ensuing years remains to be described.

References

1.1. Hansen, James R.: *Engineer in Charge: A History of the Langley Aeronautical Laboratory,* 1917–1958. NASA SP-4305, 1987.

2.1. Phillips, Hewitt: What Can Be Learned From Paper Airplanes. *Soar Tech J. Radio Control. Soaring,* no. 9, p. 109, Nov. 1992.

2.2. Phillips, Hewitt: History of the Jordan—Traveler Junior Aviation League. *Twenty-Second Annual Report of the National Free Flight Society Symposium.* George Xenakis, ed., 1989, pp. 68–72.

2.3. Phillips, Hewitt: Building a Wind Tunnel—An Educational Experience. *Nineteenth Annual Symposium of the National Free Flight Society,* 1986, pp. 73–79.

2.4. Phillips, William H.; and Levy, Charles N.: Effect of Pressure Gradients on the Boundary Layer Along a Flat Plate. Thesis, Massachusetts Institute of Technology, 1939.

2.5. Phillips, W. H.: *Effect of Structural Flexibility on the Design of Vibration-Isolating Mounts for Aircraft Engines.* NASA TM-85725, 1984.

2.6. Phillips, W. H.; and Rauscher, Manfred: Propulsive Effects of Radiator and Exhaust Ducting. *J. Aeronaut. Sci.,* vol. 8, no. 4, Feb. 1941, pp. 167–174.

2.7. Phillips, Hewitt; and Tyler, Bill: Cutting Down the Drag. *Twenty-Seventh Annual Symposium of the National Free Flight Society,* 1994, pp. 14–18.

3.1. Phillips, William H.: Recollections of the Langley Memorial Aero Lab in the Forties. A.A.H.S. J., vol. 37, no. 2, 1992, pp. 116–127.

4.1. Bryan, G. H.: *Stability in Aviation.* MacMillan, 1911.

4.2. Soulé, H. A.: *Preliminary Investigation of the Flying Qualities of Airplanes.* NACA TR-700, 1940.

4.3. Gilruth, R. R.: *Requirements for Satisfactory Flying Qualities of Airplanes.* NACA Rep. 755, 1943. (Supersedes NACA ARC, Apr. 1941.)

4.4. Phillips, William H.; and Vensel, J. R.: *Measurements of the Flying Qualities of a Curtiss P-40 Airplane.* AC No. 39-160, NACA CMR, Army Air Corps, May 1941.

4.5.   Nissen, J. M.; and Phillips, W. H.: *Measurements of the Flying Qualities of a Hawker Hurricane Airplane.* NACA WR L-565. Apr. 1942. (Formerly NACA MR, Apr. 1942.)

4.6.   Phillips, W. H.: Flying Qualities From Early Airplanes to the Space Shuttle. *AIAA J. Guid., Control, & Dyn.*, vol. 12, no. 4, 1989, pp. 449–459.

4.7.   Phillips, William H.: *A Flight Investigation of Short-Period Longitudinal Oscillations of an Airplane With Free Elevator.* NACA WR L-444, 1942. (Formerly NACA ARR, May 1942.)

4.8.   Kleckner, Harold F.: Flight Tests of an All-Movable Vertical Tail on the Fairchild XR2K-1 Airplane. NACA ACR No. 3F26, 1943.

4.9.   Phillips, W. H.; Crane, H. L.; and Hunter, P. A.: *Effect of Lateral Shift of Center of Gravity on Rudder Deflection Required for Trim.* NACA WR L-92, 1944. (Formerly NACA RB L4I06.)

4.10. Phillips, William H.: *Appreciation and Prediction of Flying Qualities.* NACA Rep. 927, 1949. (Supersedes NACA TN 1670.)

4.11. Gilruth, R. R.; and White, M. D.: *Analysis and Prediction of Longitudinal Stability of Airplanes.* NACA Rep. 711, 1941.

5.1.   Bode, Hendrik Wade: *Network Analysis and Feedback Amplifier Design.* D. Van Nostrand Co., Inc., 1945.

5.2.   Bush, Vannevar: *Operational Circuit Analysis.* John Wiley & Sons, Inc., 1937.

5.3.   Jones, R. T.: *A Simplified Application of the Method of Operators to the Calculation of Disturbed Motions of an Airplane.* NACA-TR-560, Jan. 1937.

5.4.   Gardner, Murray F. and Barnes, John L.: *Transients in Linear Systems Studied by the Laplace Transformation.* Lumped-Constant Systems, Volume I, John Wiley & Sons, Inc., 1942.

5.5.   Minorsky, Nicolai: *Introduction to Non-Linear Mechanics—Topological Methods, Analytical Methods, Non-Linear Resonance, Relaxation Oscillations.* Ann Arbor: J. W. Edwards, 1947.

5.6.   Weiss, Herbert K.: *Dynamics of Constant-Speed Propellers.* J. Aeron. Sci., vol. 70, pp. 58-67, Feb. 1943.

5.7.   Lin, Shih-Nge: A Mathematical Study of the Controlled Motion of Airplanes. Thesis, Volume 1, Massachusetts Inst. of Technology, 1939.

5.8.   Mueller, R. K.: The Graphical Solution of Stability Problems. *J. Aeronaut. Sci.*, vol. 4, no. 8, June 1937, pp. 324–331.

5.9.   Breuhaus, Waldemar O.: *Resumé of the Time Vector Method as a Means for Analyzing Aircraft Stability Problems.* WADC Tech. Rep. 52-299, U.S. Air Force, Nov. 1952.

5.10. Larrabee, E. E.: *Application of the Time Vector Method to the Analysis of Flight Test Lateral Oscillation Data.* C.A.L. FRM No. 189, Sept 1953.

5.11. Klawans, Bernard B.: *A Simple Method for Calculating the Characteristics of the Dutch Roll Motion of an Airplane.* NACA TN 3754, 1956.

6.1.   Phillips, William H.: *Application of Spring Tabs to Elevator Controls.* NACA Rep. 797, 1944. (Formerly included in NACA ARR L4H28 and NACA RB L5A13.)

6.2. Phillips, William H.; and Thompson, Robert F.: *Investigation by the Transonic-Bump Method of a 35 Degree Sweptback Semispan Model Equipped With a Flap Operated by a Series of Servovanes Located Ahead of and Geared to the Flap.* NACA TN 3689, 1956.

7.1. Phillips, William H.: *Preliminary Measurements of Flying Qualities of the Japanese Mitsubishi 00 Pursuit Airplane (U.S. Navy Designation Zero-2).* NACA Mr. Bu. Aero, May 1943.

8.1. Phillips, William H.; and Cheatham, Donald C.: *Ability of Pilots to Control Simulated Short-Period Yawing Oscillations.* NACA RM L50D06, 1950.

8.2. Phillips, William H.: *Effect of Steady Rolling on Longitudinal and Directional Stability.* NACA TN 1627, 1948.

8.3. Sternfield, Leonard: *Effect of Product of Inertia on Lateral Stability.* NACA TN 1193, 1947.

8.4. Boucher, Robert W.; Rich, Drexel A.; Crane, Harold L.; and Matheny, Cloyce E.: *A Method for Measuring the Product in Inertia and the Inclination of the Principal Longitudinal Axis of Inertia of an Airplane.* NACA TN 3084, 1954.

8.5. Phillips, W. H.: *Stability of a Body Stabilized by Fins and Suspended From an Airplane.* NACA WR L-28, 1944. (Formerly NACA ARR L4D18.)

8.6. Rayleigh, John William Strutt, (Lord): *The Theory of Sound.* Dover Publ., 1945.

8.7. Phillips, William H.: *Theoretical Analysis of Oscillations of a Towed Cable.* NACA TN 1796, 1949.

8.8. Billing, Heinz: *Oscillations excited in the cable of a bomb towed behind an aircraft.* Tech. Trans, TT-88. Nat. Reseach Council of Canada, May 3, 1949.

9.1. Theodorsen, Theodore: *General Theory of Aerodynamic Instability and the Mechanism of Flutter.* NACA Rep. 496, 1935.

9.2. Birnbaum, W.: The Uniplanar Problem of the Flapping Wing. *Z.f.a.MM*, vol. 4, no. 4, Aug. 1924, pp. 277–292.

9.3. Wagner, Herbert: Über die Entstehung des dynamischen Auftriebes von Tragflügeln. (The Origin of Dynamic Lift on Airplane Wings.) *Z.f.a.M.M.*, Bd. 5, Heft 1, Feb. 1925, pp. 17–35.

9.4. Glauert, H.: *Force and Moment on an Oscillating Aerofoil.* British R&M 1242, 1929.

9.5. Walker, P. B.: *Growth of Circulation About a Wing and an Apparatus for Measuring Fluid Motion.* British R&M 1402, 1931.

9.6. Smilg, Benjamin: The Instability of Pitching Oscillations of an Airfoil in Subsonic Incompressible Potential Flow. *J. Aeronaut. Sci.*, vol. 16, no. 11, Nov. 1949, pp. 691–696.

9.7. Pinsker, W. J. G.: *Note on the Dynamic Stability of Aircraft at High-Subsonic Speeds When Considering Unsteady Flow.* ARC R&M 2904, 1950.

9.8. Jones, Robert T.: *The Unsteady Lift of a Wing of Finite Aspect Ratio.* NACA Rep. 681, 1940.

10.1.   Phillips, William H.; Kuehnel, Helmut A.; and Whitten, James B.: *Flight Investigation of the Effectiveness of an Automatic Aileron Trim Control Device for Personal Airplanes.* NACA Rep. 1304, 1957. (Supersedes NACA TN 3637.)

11.1.   Phillips, William H.; Brown, B. Porter; and Matthews, James T., Jr.: *Review and Investigation of Unsatisfactory Control Characteristics Involving Instability of Pilot-Airplane Combination and Methods for Predicting These Difficulties From Ground Tests.* NACA TN 4064, 1957.

11.2.   McRuer, Duane: *Human Dynamics and Pilot-Induced Oscillations.* Twenty-Second Minta Martin Lecture, MIT Dept. of Aeronautics and Astronautics, Dec. 2, 1992.

11.3.   Koppen, O. C.: Airplane Stability and Control from a Designer's Point of View. *J. Aeronautical Sciences,* Feb. 1940, pp. 135–140.

11.4.   Weiss, Herbert K.: *Theory of Automatic Control of Airplanes.* NACA TN 700, 1939.

11.5.   Tustin, A.: The Nature of the Operator's Response in Manual Control and Its Implications for Controller Design. *J. Inst. Electr. Eng.,* vol. 94, pt. IIA, no. 2, 1947, pp. 190–202.

11.6.   Gustafson, F. B.; and O'Sullivan, William J., Jr.: *The Effect of High Wing Loading on Landing Technique and Distance, With Experimental Data for the B-26 Airplane.* NACA WR L-160, 1945. (Formerly NACA ARR L4K07, Jan. 1945.)

11.7.   Von Doenhoff, Albert E.; and Jones, George W., Jr.: *An Analysis of the Power-Off Landing Maneuver in Terms of the Capabilities of the Pilot and the Aerodynamic Characteristics of the Airplane.* NACA TN 2967, 1953.

11.8.   Von Doenhoff, Albert E.; and Hallissy, Joseph M., Jr.: *Systems Using Solar Energy for Auxiliary Space Vehicle Power.* Rep. No. 59-40, Inst. Aeronaut. Sci., Jan. 1959.

12.1.   Einstein, A.: On the Movement of Small Particles Suspended in Stationary Liquids Required by the Molecular-Kinetic Theory of Heat. *Annalen der Physik,* vol. 17, 1905, pp. 549–560.

12.2.   Taylor, G. I.: The Statistical Theory of Isotrophic Turbulence *J. of Aero. Sci.,* vol. 4, no. 8, Jun. 1937, pp. 311–314.

12.3.   von Kármán, T.: The Fundamentals of the Statistical Theory of Turbulence. *J. of Aero. Sci.,* vol. 4, no. 4, Feb. 1937, p. 135.

12.4.   Heisenberg, W.: On the Statistical Theory of Turbulence. Trans. of Zur Statistischen Theorie der Turbulnez. *Zeitschrift für Physik,* vol. 124, 1948. NACA TM 1431, 1958.

12.5.   Crane, Harold L.; and Chilton, Robert G.: *Measurements of Atmospheric Turbulence Over a Wide Range of Wavelength for One Meteorological Condition.* NACA TN 3702, 1956.

13.1.   Hirsch, René: *Recherches théoriques et expérimentales sur les moyens de soustraire un avion aux accélérations que peuvent engendrer les perturbations atmospheriques.* Publ. No. 138, Publ. Sci. et Tech. du Ministere de l'Air (Paris), 1938

13.2.   Phillips, William H.; and Kraft, Christopher C., Jr.: *Theoretical Study of Some Methods for Increasing the Smoothness of Flight Through Rough Air.* NACA TN 2416, 1951.

13.3. Cowley, W.L. and Glauert, H.: The Effect of the Lag of the Downwash on the Longitudinal Stability of an Aeroplane and on the Rotary Derivative. Mq. R. & M. No, 718, British A.R.C., 1921.

13.4. Kraft, Christopher C., Jr.: *Initial Results of a Flight Investigation of a Gust-Alleviation System.* NACA TN 3612, 1956.

13.5. Hunter, Paul A.; Kraft, Christopher C., Jr.; and Alford, William L.: *A Flight Investigation of an Automatic Gust-Alleviation System in a Transport Airplane.* NASA TN D-532, 1961.

13.6. Cooney, T. V.; and Schott, Russell L.: *Initial Results of a Flight Investigation of the Wing and Tail Loads on an Airplane Equipped With a Vane Controlled Gust-Alleviation System.* NACA TN 3746, 1956.

13.7. Schott, Russell L.; and Hamer, Harold A.: *Flight Investigation of Some Effects of a Vane-Controlled Gust-Alleviation System on the Wing and Tail Loads of a Transport Airplane.* NASA TN D-643, 1961.

13.8. Barker, L. Keith; and Sparrow, Gene W.: *Analysis of Effects of Spanwise Variations of Gust Velocity on a Vane-Controlled Gust-Alleviation System.* NASA TN D-6126, 1971.

13.9. Barker, L. Keith: *Effects of Spanwise Variation of Gust Velocity on Alleviation System Designed for Uniform Gust Velocity Across Span.* NASA TN D-6346, 1971.

13.10. Phillips, William H.: Gust Alleviation. *Performance and Dynamics of Aerospace Vehicles,* NASA SP-258, 1971.

13.11. Hirsch, René: Études et Essais d'un Avion Absorbeur de Rafales. *Docaéro,* no. 42, Jan. 1957.

13.12. Hirsh, R.: L'Absorption des Rafales sur Avions et Résultats des Essais en Vol d'un Appareil Expérimental. *Doc-Air-Espace,* no. 105, July 1967, pp. 41–56.

13.13. Burris, P. M.; and Bender, M. A.: *Aircraft Load Alleviation and Mode Stabilization (LAMS).* AFFDL-TR-68-158, U.S. Air Force, Apr. 1969.

13.14. Rynaski, E. G.; Andrisani, D., II; and Weingarten, N.: *Active Control for the Total-In-Flight Simulator (ACTIFS).* NASA CR-3118, 1979.

14.1. Ribner, H. S.: Spectral Theory of Buffeting and Gust Response: Unification and Extension. *J. Aeronautical Sciences,* vol. 23, no. 12, Dec. 1956.

14.2. Etkin, Bernard: *Dynamics of Flight.* John Wiley & Sons, 1959.

14.3. Diederich, Franklin W.: *The Response of an Airplane to Random Atmospheric Disturbances.* NACA Rep. 1345, 1958.

14.4. Eggleston, John M.; and Diederich, Franklin W.: *Theoretical Calculation of the Power Spectra of the Rolling and Yawing Moments on a Wing in Random Turbulence.* NACA Rep. 1321, 1957. (Supersedes NACA TN 3864.)

14.5. Eggleston, John M.; and Phillips, William H.: *The Lateral Response of Airplanes to Random Atmospheric Turbulence.* NASA TR R-74, 1960. (Supersedes NACA TN 3954 by Eggleston and TN 4196 by Eggleston and Phillips.)

14.6. Eggleston, John M.; and Phillips, William H.: *A Method for the Calculation of the Lateral Response of Airplanes to Random Turbulence.* NACA TN 4196, 1958.

15.1. National Safety Board Report. *Aviat. Week & Space Technol.*, 1968.Part 1—Turbulence Cited in BAC 111 Crash, June 10, 1968. Part 2—Low-Level Turbulence Study Urged. June 17, 1968.

15.2. Duncan, W. J.: *The Representation of Aircraft Wings, Tails and Fuselages by Semi-Rigid Structures in Dynamic and Static Problems.* British R&M 1904. 1943.

15.3. Adams, James J.; and Whitten, James B.: *Flying Qualities of a High-Performance Personal-Owner Airplane.* NACA RM SL51F18, CAA, 1951.

15.4. U.S. Dept. of Transportation, Transportation Systems Center: Task Force Report, V-Tail Bonanza Investigation. Prepared for the FAA, Jan. 1986.

16.1. Phillips, William H., compl.: *NACA Flight Research on the Vought F8U-1 Airplane,* Dec. 1981. (Available from NASA Langley Research Center, Hampton, VA.)

16.2. Kraft, Christopher C., Jr.; McLaughlin, Milton D.; White, Jack A.; and Champine, Robert A.: *Flight Measurements of Some of the Flying Qualities and Stability Derivatives of a Supersonic Fighter Airplane.* NASA Memo Rep. 10-7-58L, 1958.

18.1. Katzoff, S.: *Clarity in Technical Reporting.* NASA SP-7010, 1964.

19.1. Hansen, James R.: *Spaceflight Revolution—NASA Langley Research Center From Sputnik to Apollo.* NASA SP-4308, 1995.

19.2. Morin, Relman: Expert Says Sputnik Won The Cold War. *Newport News (VA) Daily Press,* Jan. 7, 1958, p. 1.

Early Autobiography

This autobiography was written as a class assignment when entering the tenth grade English class at the age of fourteen.

Hewitt Phillips
Grade 10   Eng 2A
Sept. 16, 1932

My Autobiography

I was born at Port Sunlight, Cheshire, England, on May 31st, 1918. Port Sunlight is a model village, built around, Lever Brothers soap works near Birkenhead and Liverpool. There are only two model villages in England, Port Sunlight, and the "chocolate village" of Bournville. Therefore my earliest surroundings were very picturesque. My father was born in Durham and my mother in Chester. My Father worked at the Lever Brothers' firm. Our house overlooked the River Mersey, so during my babyhood I liked to watch the large steamers come up the river. These always interested me. I also enjoyed crossing over to Liverpool with my parents in the Ferry boats, a sail of about 15 minutes. At other times we took walks around the pretty village of Port Sunlight and further into the beautiful countryside of Cheshire. Going to Raby Mere, a lake resort, was one of our favorite walks. I had Grandparents at Warrington, a distance of 18 miles away, and I often went there by train.

At the age of two, when I was just becoming interested in all these things, my father's business necessitated his coming to America to the Crane Brothers works at Cambridge, therefore we all moved to this country, travelling on the White Star liner "Celtic." We then came to Boston and made our home in Watertown. I lived here three years. I was now five years old and had been in America three years, when my parents took a trip to England. This was the summer of 1903 and we stayed from April to September. We crossed on the Cunard steamer "Ivernia" and returned on the "Scythia." While in England I visited Port Sunlight, Liverpool, Warrington, Middlepool, and Newcastle-on-Tyne. I was now old enough to take a great interest in these places and I can still remember some of them. As this trip was made largely on boats and trains, I was very interested in them, and I liked to draw and paint pictures of them. When we returned to America we decided to leave Watertown and to reside in Belmont, chiefly because there was no school near enough for me to attend, so we bought a house at Spinney Terrace where we have lived ever since.

I entered the Roger Wellington School and later attended the Homer, Roger Wellington again, Junior High, and have now entered the Belmont High School.

As I grew older my interest in mechanical things remained. I still drew pictures of boats and trains, and I also became interested in airplanes. From the age of seven I have had a hobby of making model boats and airplanes. I make my own boats to take to the beach in Summer, and I have made many model airplanes. I joined the Jordan-Traveller Junior Aviation League in 1930 and this Spring I won the first prize in the Bellanca Scale Model contest.

I am now an American citizen as my parents became naturalized as soon as possible, that is, five years, after entering the country. I like America very much and have travelled to many places in New England.

Next year I hope to visit England again, so that I can see other parts of England that I have not yet been to, and also to reunite with my Grandparents and other relatives again

Bibliography of Reports by William Hewitt Phillips

NACA or NASA Reports

1. Phillips, William H.; and Vensel, J. R.: Measurements of the Flying Qualities of a Curtiss P-40 Airplane (AC No. 39-160). NACA CMR, Army Air Corps, May 31, 1941.

2. Phillips, William H.; and Gough, Melvin N.: Investigation of Aileron Characteristics of the Vought-Sikorsky XF4U-1 Airplane. Memorandum Report for Navy, Aug. 15, 1941.

3. Phillips, William H.; and Gough, Melvin N.: Measurements of the Longitudinal Stability and Control Characteristics of the Vought-Sikorsky XF4U-1 Airplane. Memorandum Report for Navy, Aug. 30, 1941.

4. Phillips, William H.: Effect of Spring and Gravity Moments in the Control System on the Longitudinal Stability of the Brewster XSBA-1 Airplane. NACA ARR, Apr. 1942. (Wartime Report L-263.)

5. Phillips, William H.; and Nissen, J. M. : Measurements of the Flying Qualities of a Hawker Hurricane Airplane. NACA MR for AAF, Apr. 20, 1942. (Wartime Report L-565.)

6. Phillips, William H.: A Flight Investigation of Short Period Longitudinal Oscillations of an Airplane with Free Elevator. NACA ARR, May 1942. (Wartime Report L-444.)

7. Phillips, William H.: Control-Surface Instability on High Speed Airplanes. NACA CB, June 1942.

8. Phillips, William H.: Comparison of Aileron Control Characteristics as Determined in Flight Tests of P-36, P-40, Spitfire, and Hurricane Pursuit Airplanes. NACA CB, Aug. 1942.

9. Phillips, William H.; and Vensel, J. R.: Measurements of the Flying Qualities of the Supermarine Spitfire VA Airplane. NACA ACR, Sept. 1942. (Wartime Report L-334.)

10. Phillips, William H.; and Vensel, J. R. : Stalling Characteristics of the Supermarine Spitfire VA Airplane. NACA ACR, Sept. 1942. (Wartime Report L-353.)

11. Phillips, William H.: Preliminary Measurements of Flying Qualities of the Japanese Mitsubishi 00 Pursuit Airplane. (US Navy Designation Zero-2) NACA MR Bu. Aero. May 5, 1943.

12. Phillips, William H.; and Nissen, J. M.: Flight Tests of Various Tail Modifications on the Brewster XSBA-1 Airplane. I—Measurements of Flying Qualities with Original Tail Surfaces. NACA ARR 3F07, June 1943. (Wartime Report L-612.)

13. Phillips, William H.; and Crane, Harold L.: Flight Tests of Various Tail Modifications on the Brewster XSBA-1 Airplane. II—Measurements of Flying Qualities with Tail Configuration Number Two. NACA MR, Dec. 1943. (Wartime Report L-589.)

14. Phillips, William H.: Stability of a Body Stabilized by Fins and Suspended From an Airplane. NACA ARR L4D12, Apr. 1944. (Wartime Report L-28.)

15. Phillips, William H.; Williams, W. C.; and Hoover, H. H.: Measurements of Flying Qualities of a Curtiss SB2C-1 Airplane (No. 000 14). NACA MR, Bu. Aero. Mar. 14, 1944. (Wartime Report L-571.)

16. Phillips, William H.: Application of Spring Tabs to Elevator Controls. NACA ARR L8H28, Oct. 1944. (Wartime Report L-122.)

17. Phillips, William H.; Crane, Harold L.; and Hunter, Paul A.: Effect of Lateral Shift of Center of Gravity on Rudder Deflection Required for Trim. NACA RB L4I06, Nov. 1944. (Wartime Report L-92.)

18. Phillips, William H.: Application of Spring Tabs to Elevator Controls. NACA Report No. 797, 1944.

19. Phillips, William H.: An Investigation of Additional Requirements for Satisfactory Elevator Control Characteristics. NACA MR L5G16, May 1945. (Republished as TN 1060, June 1946.)

20. Phillips, William H.: The Use of Geared Spring Tabs for Elevator Control. NACA RB L5A13, Feb. 1945. (Wartime L-30.)

21. Phillips, William H.: Effect of Steady Rolling on Longitudinal and Directional Stability. NACA 1627, June 1948.

22. Phillips, William H.: Appreciation and Prediction of Flying Qualities. NACA TN-1670, Aug. 1948. (Republished as NACA Report 927, 1949.)

23. Phillips, William H.: Theoretical Analysis of Oscillations of a Towed Cable. NACA TN 1796, Nov. 1948.

24. Phillips, William H.; and Adams, J. J.: Low-Speed Tests of a Model Simulating the Phenomenon of Control-Surface Buzz. NACA RM L50F19, Aug. 1950.

25. Phillips, William H.; and Cheatham, Donald C.: Ability of Pilots to Control Simulated Short-Period Yawing Oscillations. NACA RM L50D06, Nov. 1950.

26. Phillips, William.; and Kraft, C. C., Jr.: Theoretical Study of Some Methods for Increasing the Smoothness of Flight Through Rough Air. NACA TN 2416, July 1951.

27. Phillips, William, H.; and Thompson, R. F.: Investigation by the Transonic-Bump Method of a 35° Semispan Model Equipped With a Flap Operated by a Series of Servo-Vanes Located Ahead of and Geared to the Flap. NACA RM L51J10, Dec. 1951. (Republished as TN 3689, Apr. 1956.)

28. Phillips, William H.: Theoretical Analysis of Some Simple Types of Acceleration Restrictors. NACA TN 2574, Dec. 1951.

29. Phillips, William H.; Brown, B. Porter; and Matthews, James T., Jr.: Review and Investigation of Unsatisfactory Control Characteristics Involving Instability of Pilot-Airplane Combination and Methods for Predicting These Difficulties From Ground Tests. NACA RM L53F17a, Aug. 1953. (Republished as TN 4064, Aug. 1957.)

30. Phillips, William H.: Graphical Solution of Some Automatic-Control Problems Involving Saturation Effects With Application to Yaw Dampers for Aircraft. NACA TN 3034, Oct. 1953.

31. Phillips, W. H. ; and Kuehnel, H. A.: Theoretical Investigation of Some Discontinuous Yaw Dampers. NACA RM L54B24a, Apr. 1954.

32. Phillips, W. H. : Investigation of Effect of Reduction of Valve Friction in a Power Control System by Use of a Vibrator. NACA RM L55E18a, July 1955.

33. Phillips, W. H.; Kuehnel, H. A.; and Whitten, J. B.: Flight Investigation of the Effectiveness of an Automatic Aileron Trim Control Device For Personal Airplanes. NACA TN 3637, Apr. 1956. (Changed to NACA Report 1304.)

34. Phillips, William H.: Analysis of an Automatic Control to Prevent Rolling Divergence. NACA RM L56A04, Apr. 1956.

35. Phillips, William H.: Loads Implication of Gust-Alleviation Systems. NACA TN 4056, June 1957.

36. Eggleston, John M.; and Phillips, William H.: The Lateral Response of Airplanes to Random Atmospheric Turbulence. TR R-74, 1960. (Supersedes TN-4196.)

37. Phillips, William H.: Analysis of Effects of Interceptor Roll Performance and Maneuverability on Success of Collision Course Attacks. NASA TN D-952, Aug. 1961. (Supersedes RM L-58E27.)

38. Phillips, William H.: Research on Blunt Faced Entry Configurations at Angles of Attack Between 60° and 90°. NASA TM X-315, Sept. 1960.

39. Phillips, William H.: NASA Reports—A Discussion for Prospective Authors. Jan. 3, 1962.

40. Phillips, W. H.; Queijo, M. J.; and Adams, James J.: Langley Research Center Simulation Facilities for Manned Space Missions. NASA TM X-50390, July 1965.

41. Phillips, William H.: Propulsive Efficiency of Tail-Mounted Turbofan Engines. NASA LWP-693, Dec. 17, 1968.

42. Phillips, William H.: Gust Alleviation. In: Performance and Dynamics of Aerospace Vehicles. NASA SP-258, 1971.

43. Phillips, William H.: Study of a Control System to Alleviate Aircraft Response to Horizontal and Vertical Gusts. NASA TN D-7278, Dec. 1973.

44. Phillips, William H.; and Lichtenstein, Jacob H.: Comparison of Effects of Unsteady Lift and Spanwise Averaging in Flight Through Turbulence. NASA TM X-74028, June 1977.

45. Phillips, William H.: Ground Distance Covered During Airborne Horizontal Deceleration of an Airplane. NASA TP-1157, Apr. 1978.

46. Phillips, William H.: Simulation Study of the Oscillatory Longitudinal Motion of an Airplane at the Stall. NASA TP-1242, Aug. 1978.

47. Phillips, William H.: Altitude Response of Several Airplanes during Landing Approach. NASA TM-80186, Nov. 1979.

48. Some Design Considerations for Solar-Powered Aircraft. NASA TP-1675, June 1980.

49. Phillips, William H.: Effect of Structural Flexibility on the Design of Vibration-Isolating Mounts for Aircraft Engines. NASA TM-85725, Feb. 1984.

50. Phillips, William H.; and Grafton, Sue B.: Study of Canard Surfaces on the Shuttle Orbiter. NASA TM-89156, Sept. 1987.

51. Phillips, William H.: Determination of the Pressure Drag of Airfoils by Integration of Surface Pressures. NASA TM-102722, Oct. 1990.

Papers Other Than NACA or NASA Reports

1. Phillips, William H.; and Rauscher, M.: Propulsive Effects of Radiator and Exhaust Ducting. J. Aeronaut. Sci., vol. 8, no. 4, Feb. 1941, p.167.

2. Phillips, William H.: Flying Qualities Requirements for Personal Airplanes. NACA—Industry Conference on Personal Aircraft Research, Sept. 1946.

3. Phillips, William H.: Low-Speed Flight Investigation of an Airplane with Sweptback Wings. NACA Conference on Aerodynamic Problems of Transonic Airplane Design, Nov. 1947, pp. 145–152.

4. Phillips, William H.: Flying and Handling Qualities of Airplanes. NACA—University Conference on Aerodynamics, June 1948.

5. Phillips, William H.: Flying Qualities of Airplanes at Transonic Speeds. Talk at Test Pilot Training School, NATC, Patuxent, June 2, 1949.

6. Phillips, William H.: Analysis of Means to Increase the Smoothness of Flight Through Rough Air. NACA Conference on Some Problems of Aircraft Operation, Oct. 1950.

7. Phillips, William H.: Power Control Systems, Artificial Feel, and Acceleration Restrictors. NACA Conference on High-Speed Airplane Aerodynamics, Dec. 1951.

8. Phillips, William H.: Recent Studies of Stability and Control Characteristics of Fighter Airplanes and Their Relation to Tactical Requirements. Talk at Test Pilot Training School, NATC. Patuxent, June 13, 1952.

9. Phillips, William H.: Stability and Control Problems at Transonic and Supersonic Speeds and Problems of Aiming and Sighting Guns at Transonic Speeds. Talk for Ad Hoc Group on Airborne Fire Control Systems for Fighter Aircraft, Research and Development Board, Dec. 11, 1952.

10. Phillips, William H.: High-Speed Stability and Control Problems. Presented to Flight Test Panel of the Advisory Group for Aeronautical Research and Development of NATO, May 1953.

11. Phillips, William H.: High-Speed Flight Stability and Control Problems. Talk at Test Pilot Training School, NATC, Patuxent, June 28, 1954.

12. Phillips, William H.: Power Control Systems and Control Feel Devices for Satisfactory Airplane Handling Qualities. NACA Conference on Some Problems of Aircraft Operation, Nov. 1954.

13. Phillips, William H.: A Study of Means of Limiting the Maximum Normal Acceleration of Airplanes. NACA Conference on Automatic Stability and Control of Aircraft, Mar. 1955.

14. Phillips, William H.; and Williams, W. C.: Some Recent Research on the Handling Qualities of Airplanes. NACA Conference on Aerodynamics of High-Speed Aircraft, Nov. 1955.

15. Phillips, William H.: Similarities and Contrasts Between Ship Design and Aircraft Design-Stability and Control. Bull. Soc. Naval Arch. & Mar. Eng., vol. XI, no. 2, July 1956.

16. Phillips, William H.: Loads Implications of Gust-Alleviation Systems. NACA Conference on Aircraft Loads, Structures, and Flutter. Mar. 1957.

17. Phillips, William H.: High-Speed Stability and Control Problems. AGARD Flight Test Manual, Vol. II, Chapt. 9, Pt. I.

18. Phillips, William H.: Flight Measurements of the Stability and Controllability of Airplanes. Princeton Series, Volume VII—Experimental Methods in High-Speed Aerodynamics.

19. Phillips, William H.; Queijo, M. J.; and Adams, James J.: Langley Research Center Simulation Facilities for Manned Space Missions. Published in ASME Publication—Paper No. 63-AHGT-91. (Contributed by the Aviation and Space Division for Presentation at the Aviation and Space, Hydraulic, and Gas Turbine Conference and Products Show. Los Angeles, CA., Mar. 1963.

20. Phillips, William H.: Gust Alleviation. Lecture presented at Rensselaer Polytechnic Institute, Spring 1970.

21. Phillips, William H.: Analysis of Effect of Asymmetric Loading on Sailplane Performance in Circling Flight. Paper presented at the Symposium on the Technology and Science of Motorless Flight. Cambridge, Massachusetts (M.I.T.), Oct. 1972.

22. Phillips, William H.: Analysis and Experimental Studies of the Control of Hang Gliders. Presented at the AIAA/MIT/SSA Second International Symposium on the Technology and Science of Low-Speed and Motorless Flight, Cambridge, Massachusetts, Sept. 1974.

23. Phillips, William H.: Stability and Control of Hang Gliders. Presented at the Soc. of Auto. Eng. Meeting. Dayton, Ohio, Sept. 1976.

24. Phillips, William H.: Gyroscopic Moments on a Glider in Turing Flight. Technical Soaring, 1973.

25. Stewart, E. C.; Phillips, W. H.; and Hewes, D. E.: Gust Alleviation System to Improve the Ride Comfort of Light Airplanes. NACA Technology Utilization Program Report LAR-11711, Oct. 1975.

26. Phillips, William H.: Propulsive Effects Due to Flight Through Turbulence. Engineering Note Published in J. Aircr., vol. 12, no. 7, July 1975, pp. 624–626.

27. Phillips, W. H.: Recovery from a Vertical Dive. *Ground Skimmer*, No. 31, Aug. 1975, pp. 18–21.

28. Phillips, W. H.: More on Dive Recovery, or Lack Thereof. *Ground Skimmer*, No. 34, Nov. 1975, pp. 24–25.

29. Phillips, William H.; and Shaughnessy, John D.: Investigation of Longitudinal Control System for a Small Hydrofoil Boat. *J. of Hydro.*, vol. 10, no. 2, Apr. 1976.

30. Phillips, W. H.: Building a Wind Tunnel—An Educational Experience. 19th Annual Symposium of National Free Flight Society, 1986, pp. 73–79.

31. Phillips, William H.: Propeller Design Program. NASA Technology Utilization Program Report. LAR-13141 and LAR-12142, vol. 8, no. 4, Apr. 1985.

32. Phillips, W. H.: Research on the Energy Efficiency of Fireplaces. Contract DE-FG43-81R308099, (Work conducted at VARC for DOE. Final Report), College of William and Mary, Virginia, Associated Research Center Campus, Oct. 1982.

33. Phillips, William H.: Flying Qualities from Early Airplanes to the Space Shuttle. J. Guid., Control, & Dyn., vol. 12, no. 4, July1989, pp. 449–459.

34. Phillips, William H.: Studies of Friction Drag and Pressure Drag of Airfoils Using the Eppler Program. SAE TP-881396, Oct. 1988.

35. Phillips, William H.: Recollections of the Langley Memorial Aero Lab in the Forties. J. American Aviat. Hist. Soc., vol. 37, no. 2, Summer 1992, pp. 116–127.

36. Phillips, W. H.: Effects of Model Scale on Flight Characteristics and Design Parameters. AIAA J. Aircr., vol. 31, no. 2, Mar. 1994, pp. 454–457.

Unpublished Papers in NASA Langley Research Center Library

1. Phillips, William H.: Dynamic Longutidanal Stability of Airplanes, L.A.L. Notes for Lectures Presented in UVA Extension Course, 1956. (Unpublished, Library No. *CN-47503.)

2. Phillips, W. H.: Correlation of Data Obtained by various test methods at Transonic Speeds. Oct. 7, 1946. (Unpublished, Library No. 1111262.)

Solution of a High School Geometry Problem

This proof was completed in a high school course on Euclidean geometry. The teacher had no expectation that anyone could solve it, but I managed to give a proof of the proposition in 45 steps.

Hewitt Phillips   *Excellent!*   Pl. Geom B

Pg 118, ex 5 Given: equilateral $\triangle ABC$, with altitude CD and from any point O inside the $\triangle$ 3 $\perp$s drawn to the sides at E, F, and G on AB, BC, and CA resp.

To prove: $OE + OF + OG = DC$

Construction: extend AC, extend OF to K, so that OK = OG. Through K draw KH || to BC intersecting AC (extended) at H. draw OH. from C draw $C \perp$ to OH. draw OF & HF || to AC

Proof.

1. HO = HO	1. by identity
2. GO = OK	2. by const
3. HK = HK, BC	3. by const
4. OF $\perp$ to BC	4. given
5. $\therefore$ OK is $\perp$ to KH	5. if a st. line is $\perp$ to 1 of 2 || lines, it is $\perp$ to the other
6. OG $\perp$ to GH	6. given
7. Hence, $\triangle OGH \cong \triangle OKH$	7. H.S. = H.S.
8. $\angle B = 60°$	8. $\angle$s of an equil. $\triangle = 60°$ each
9. $\angle OEB$ and $\angle OFB$ each = 90°	9. given
10. $\angle EOF + \angle OFB + \angle B + \angle BEO = 360°$	10. sum of 4 $\angle$s of a quad. = 360
11. $\therefore \angle EOF = 120°$	11. subtr. axi.
12. in the same way $\angle FOG = 120°$	
13. $\angle HOK = \angle HOG$, or $\angle GOK$ is bisected	13. c.p.c.t.o.
14. $\therefore \angle HOK = 60°$	14. halves of equals are =

Statements	Reasons
15. $\angle HOK + \angle KOE$, or $\angle HOE = 180°$ or a st. $\angle$	15. addn. ax.
16. CF is $\parallel$ to HJ	16. by const.
17. CH is $\parallel$ to FJ	17.
18. $\therefore CHJF$ is a $\square$	18. a quad with the opp. sides $\parallel$ is a parallelogram
19. $\therefore CH = JF$	19. opp. sides of a $\square$ are $=$
20. $\angle C1H$ and JKF are rt. $\angle s$	20. by const. (and one proved)
21. $CD = CD$	21. by identity
22. $CA = BC$	22. sides of an equi. $\triangle$ are $=$
23. $\angle CDA = \angle CDB$	23. all rt. $\angle s$ are $=$
24. hence, $\triangle CDA \cong \triangle CDB$	24. $H's = H S$
25. $\therefore \angle ACD = \angle BCD$	25. c p c t e
26. $\therefore \angle ACD = 30°$	26. halves of equals are $=$
27. $1E$ and CD are each $\perp$ AB	27. given
28. $\therefore CD$ is $\parallel$ to $1E$	28. lines $\perp$ to the same line are $\parallel$
29. $\therefore \angle CHE = \angle ACD$	29. corr. $\angle s$ of $\parallel$ lines are $=$
30. $\therefore \angle CHE = 30°$	30. by subsln.
31. $\angle ACB = 60°$	31. $\angle s$ of an equi $\triangle = 60°$
32. $\therefore \angle QFJ = 60°$	32. alt. int. $\angle s$ of $\parallel$ lines are $=$
33. $\angle CFK = 90°$	33. given
34. $\therefore JFK = 30°$	34. subln. ax.
35. $\therefore \angle CHE = \angle JFK$	35. quant. $=$ to the same quant. are $=$
36. hence $\triangle C1H \cong \triangle FKJ$	36. $H.A = H.A$
37. $\triangle GOH$ and KOH are $30-60$ rt. $\triangle$	37. proved
38. $\therefore GO = \frac{1}{2}HO$, and $OK = \frac{1}{2}HO$, or $GO + OK = HO$	38. in a $30-60$ rt. $\triangle$ the side opp. the $30°$ angle $= \frac{1}{2}$ the hypotenuse
39. $FK = H1$	39. c p c t e.
40. $\therefore FO + OG = 1O$	40. subln. ax.
41. $1C$ is $\parallel$ to DE	41. lines $\perp$ to the same line are $\parallel$
42. $\therefore DE1C$ is a $\square$	42. a $\square$ is a figure whose opp. sides are $\parallel$
43. $1E = CD$	43. opp. sides of a $\square$ are $=$

Solution of Quartic Equation in Terms of Two Parameters by Use of Charts

Consider the quartic equation

$$a'x^4 + b'x^3 + c'x^2 + d'x + e' = 0$$

The coefficient of the first term may be eliminated by dividing through by it, yielding

$$x^4 + \left(\frac{b'}{a'}\right)x^3 + \left(\frac{c'}{a'}\right)x^2 + \frac{d'}{a'}x + \frac{e'}{a'} = 0$$

Rewrite this equation as

$$x^4 + bx^3 + cx^2 + dx + e = 0$$

Let $x = z - \dfrac{b}{4}$

Hence $z = x + \dfrac{b}{4}$

Then

$$z^4 + lz^2 + mz + n = 0$$

where

$$l = c - \frac{3}{8}b^2$$

$$m = \frac{b^3}{8} - \frac{bc}{2} + d$$

$$n = e + \frac{b^2c}{16} - \frac{db}{4} - \frac{3b^4}{256}$$

Divide through by n to change constant coefficient to 1.

$$\frac{z^4}{n} + \frac{lz^2}{n} + \frac{mz}{n} + 1 = 0$$

Let

$$\lambda = \frac{z}{n^{1/4}}$$

Then

$$\lambda^4 + \alpha_2 \lambda^2 + \alpha_1 \lambda + 1 = 0$$

Where

$$\alpha_2 = \frac{l}{n^{1/2}}$$

$$\alpha_1 = \frac{m}{n^{3/4}}$$

The last quartic may be factored

$$\left(\lambda^2 + a_1 \lambda + b_1\right)\left(\lambda^2 - a_1 \lambda + \frac{1}{b_1}\right) = 0$$

Note that

$$\alpha_2 = \frac{1}{b_1} + b_1 - a_1^2$$

$$\alpha_1 = a_1\left(\frac{1}{b_1} - b_1\right)$$

The last two formulas are plotted in the accompanying chart, allowing b_1 to be obtained in terms of α_2 and α_1. Then

$$a_1 = \frac{\alpha_1}{\left(\dfrac{1}{b_1} - b_1\right)}$$

The quadratic factors may be readily solved to obtain the roots λ_1 and λ_2. Then, working back through the preceding derivation

$$z = \lambda_{(1,2)} n^{1/4}$$

$$x = z - \frac{b}{4}$$

FIGURE A4.1. Chart to obtain the parameter b_1 **in terms of** α_1 **and** α_2 **as given in preceding derivation.**

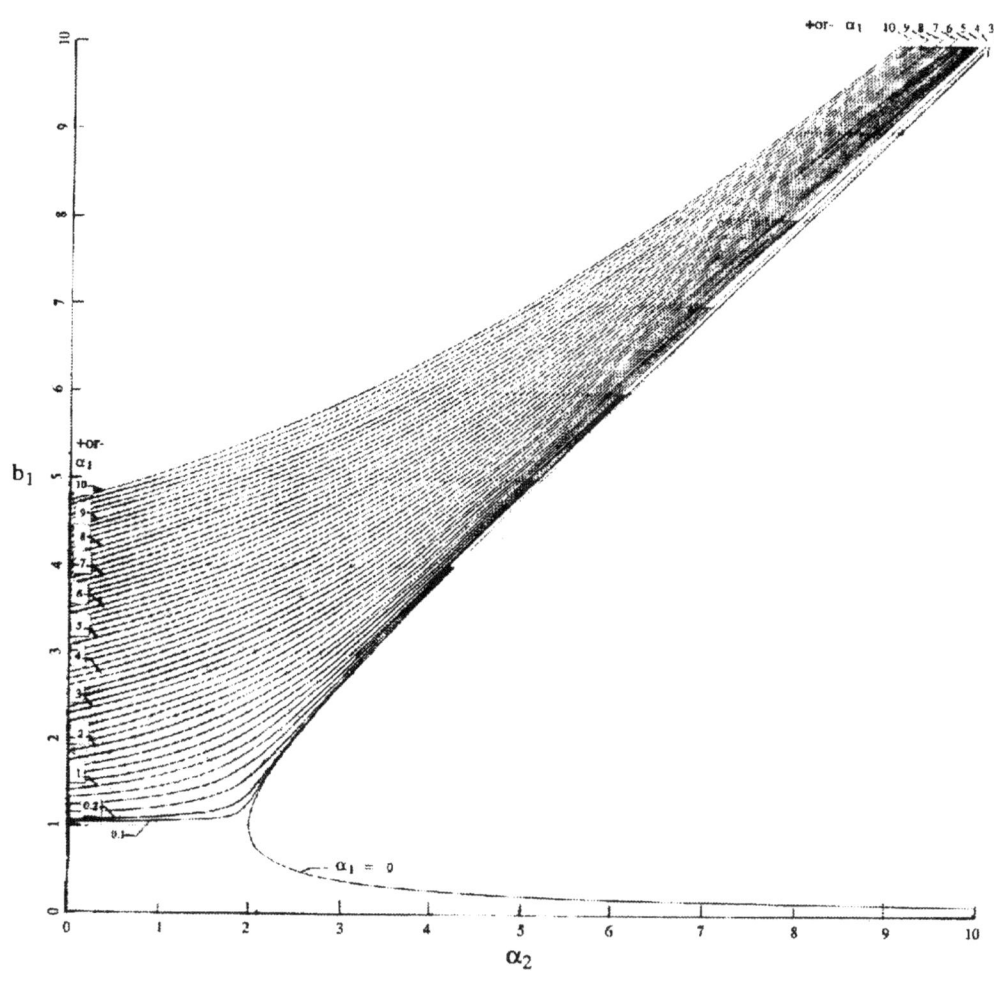

Index

About the Author

William Hewitt Phillips was born in Port Sunlight, Merseyside, England, and came to the United States with his parents at the age of 2. He was educated in the Belmont, Massachusetts, public schools and studied aeronautical engineering at MIT where he obtained his S.B. degree in 1939 and his S.M. degree in 1940. His entire professional career has been spent with the NACA, later NASA, at Langley Research Center in Hampton, Virginia. Langley Research Center is the original government center for aeronautical research in the United States. On entering duty in July 1940, he was assigned to the Flight Research Division. He specialized in the study of flying qualities and stability and control of airplanes. His duties included studies to improve the flying qualities of many World War II military airplanes. After the war, he was involved in research on the development of jet-powered fighter airplanes, supersonic airplanes, stability augmentation and its effect on human pilot control, automatic control, gust alleviation, and aeroelastic effects. This book covers his career to roughly the start of the space program in 1958, though some aeronautical work beyond this date is included. After the start of the space program, he became chief of the Space Mechanics Division and supervised 80 to 90 people in the areas of space rendezvous, navigation, and lunar landing. As a part of its responsibility to the space program, this division developed simulators for the Gemini and Apollo programs. He developed the Lunar Landing Facility that was used for training astronauts in landing on the moon. His work also included consultation and analysis in the development of the Space Shuttle. Later work included supervising studies of effects of turbulence and of application of control theory and contributing to the development of the Differential Maneuvering Simulator, a facility used for studies of air combat. He retired from government service in February 1979, but has continued in the position of Distinguished Research Associate, during which he performed original research on solar-powered aircraft, propellers, airfoil design, and wind-tunnel studies of the use of canard surfaces for the Space Shuttle. He served as a consultant on studies of flight dynamics and control. He has received numerous awards throughout his career, including the IAS Lawrence Sperry Award for aeronautics in 1944 and the President's Award for Distinguished Federal Civilian Service in 1979. At the age of 80 in 1998, he continues to be active as a Distinguished Research Associate.

Phillips married Viola Ohler in 1947 when she was head of the Editorial Office at Langley. They had three children, Frederick H., Robert O., and Alice B. Phillips All are now married. Frederick, whose wife is Joanne, is a financial consultant. Robert and wife Cheryl have three children: Tyler, 19; Ross, 16; and Jocelyn, 14. Robert works at the The Volpe Center, Cambridge, Massachusetts. Alice and husband Thomas Check have three children: Candice, 12; Nolan, 10; and Aubree, 8. Alice formerly worked for robotics firms and is now a homemaker.

www.ingramcontent.com/pod-product-compliance
Lightning Source LLC
Chambersburg PA
CBHW081441170526
45166CB00008B/2269